an explanation of redshifts
in a static Universe

Tired Light

lyndon ashmore

Cover photograph: Courtesy NASA/JPL-Caltech.
http://photojournal.jpl.nasa.gov/catalog/PIA20917

www.tiredlight.org

*All rights reserved. No part of this publication may be reproduced or transmitted in any form or by any means, electronic or mechanical,
Including photocopy, recording, or any information storage and retrieval system, without permission in writing from the author or publisher*

© lyndonashmore 2016

In the same series

Big Bang Blasted!

The story of the expanding Universe and how it was shown to be wrong!

- **ISBN-10:** 1419639226

- **ISBN-13:** 978-1419639227

Lyndon ashmore

Available from Amazon and Kindle

Chapter 1

Once upon a time

On top of Mars hill, Flagstaff, Arizona, Vesto Slipher, a fastidious little man worked quietly in the dark. The room was dimly lit by the coloured glow of light emitted from various gases held in discharge tubes and providing the emission spectra used as laboratory references. Working alone, immaculate in his suit and tie and with telescope pointing to the heavens, he worked his magic and collected spectra from distant galaxies. Armed with spectrograms from forty or so distant galaxies, he then presented his findings to the Astronomical Society of America at the campus of Northwestern University.

At the end of his presentation everyone in the room rose as one and cheered; something never before seen at a scientific meeting. What Slipher had shown was that characteristic lines in the spectra of light from distant galaxies had a longer wavelength on arrival at Earth than those same lines when measured against laboratory standards. That is, these characteristic emission lines had been shifted towards the red end of the spectrum – a phenomenon known as '*redshift*.' Slipher had also realised that the dimmer the galaxy, the greater the increase in wavelength - but what did that mean? He had no way of determining the distance to these galaxies. Usually in astronomy 'dimmer' means that the object is further away. On the other hand it just may mean that the objects are well, dimmer. As a result Slipher could not show a relationship between the redshift and distance.

At the back of the hall and standing and cheering along with everyone else was one Edwin Hubble. Hubble would later go on to work at the Mount Wilson Observatory where the greatest telescope of the age stood and using this he could not only measure the increase in wavelength but also measure the corresponding distance. Hubble looked for a special type of star in these galaxies; ones that varied in brightness in a regular pattern. These stars or *'Cepheid variables'* have a link between their Luminosity or intrinsic brightness and their time period i.e. more luminous stars take longer to go through one cycle of bright, dim bright than less luminous ones.

A simple way of determining distance is to find a Cepheid variable in a nearby galaxy and measure its period. Then turn to your more distant galaxy and find a Cepheid variable with the same period as the first one. Since they have the same period they are of equal luminosity. Now apparent brightness or how bright it appears to us on Earth depends on how far away the object is and obeys the inverse square law. So if this Cepheid variable in the more distant galaxy appears only one quarter as bright as the nearer one then it is twice as far away. If it appears one ninth as bright then it is three times further away. What Hubble found was that a galaxy twice as far away had twice the redshift and a galaxy three times farther away had three times the redshift.

Shifts in wavelength in the light from other heavenly bodies were nothing new as they can be caused by the relative motion of observer and/or source and explained by the Doppler Effect. Think of a duck swimming along. The waves emitted in the forward direction are squashed together whilst those emitted in a backward direction are stretched apart. Similarly, a celestial body moving towards us has its light waves squashed together to give a shorter wavelength shifting the characteristic emission lines to a shorter wavelength nearer the blue end of the spectrum, '*blueshift.*' One moving away has its light waves stretched to give a longer wavelength shifting the characteristic emission lines to a longer wavelength nearer the red end of the spectrum, '*redshift.*' The greater the relative speed of the source/observer then the greater the shift in wavelength and they are related by a mathematical formula. If λ is the wavelength of the characteristic emission line as measured in the laboratory, v the relative velocity of source /observer, c the speed of light in a vacuum and $\Delta\lambda$ the shift in wavelength (the difference between the wavelength as measured in the laboratory and the wavelength of the same line in the spectra of the distant galaxy) then we have; $v = c(\Delta\lambda/\lambda)$. This is how we can tell how fast stars in our own galaxy are moving (known as their '*peculiar velocity*') and even the speed at which objects such as our Sun are rotating as it spins one side moves towards us giving a blueshift whilst the other side moves away giving a redshift.

Consequently, it is not surprising that astrophysicists such as Hubble measured the redshift and consequently calculated a Doppler velocity. The problem was that these velocities of the distant galaxies were far, far greater than anything that had been measured before. Not just that, but once the galaxies were sufficiently far away that gravitational effects between them were insignificant the velocities were always recessional. They were all whizzing away from each other and the speed with which they did this was proportional their distance apart.

Since they had always calculated Doppler velocities in this way no-one stopped to think 'maybe these are something else?'

It was just assumed that these redshifts too were caused by velocities and so the next question was, 'how is it in this great and wonderful Universe a galaxy twice as far away just happens to be moving away from us twice as fast?' Isn't that a little bit too good to be true?

Well, not if we invoke God!

In comes Georges Lemaitre, a Belgian Jesuit Priest who was also a pretty good astronomer and a professor of Physics one must add who said that if we have a single point of creation, our Big Bang so to speak, then all bits and pieces would be flung out randomly at differing speeds. Since our Universe now started at a certain point in time, all the pieces of the jigsaw that is our Universe will have been travelling for the same length of time thus those that were, by chance, flung out twice as fast from the '*BANG*' will have travelled twice as far – hence our relationship between recessional velocity and redshift.

This seemed to fit in with Einstein's general theory of relativity if his cosmological constant was removed. The cosmological constant is what is mathematically termed a '*fiddle factor*,' and introduced solely to make it work. Einstein favoured a static Universe and so introduced the fiddle factor to make the equations do what he liked. This fiddle factor has been in out, in, out and shake it all about for the last one hundred years and the last time I looked it was back in!

So all was well with the Universe, science was happy as theory as per general relativity matched observation results with an expanding Universe and Religion was happy as it now had a point of creation.

Hubble and his team continued to work on redshifts and once a week he would call a meeting in the front room of his home to catch up on their research. With his team gathered round a chalkboard they were discussing these recessional velocities when one of his team, Fritz Zwicky, held up his hand and said,' wait a minute, how do we know these redshifts are velocities?'

Now Fritz was one to be listened to as he is described as a bit of a character. Mercurial some say, others quote the way he is said to have described his students as 'spherical bastards,' on the basis that it didn't matter from which angle you viewed them they would still be bastards!

Fritz went on to say that maybe these redshifts were not

due to the Doppler Effect at all but due to the photons of light losing energy as they travelled through space. Light is not a continuous wave but is made up of photons each with an amount of energy dependent on their frequency. '*Quantum,*' means '*the smallest possible amount,*' as in the excellent James Bond movie '*A Quantum of Solace,*' and each photon carries the smallest amount of energy possible for light at that frequency. If E is the energy of the photon, h the plank constant and f the frequency of the light then the energy of the photon is given by $E = hf$. So Zwicky proposed that since these photons had been battling their way across the Universe for billions of years; maybe they lost a little bit of energy as they travelled due to interactions with objects they met or passed in the intervening space? If the energy was reduced the wavelength would increase since $c = f\lambda$. The relationship between distance and redshift now becomes, 'photons of light from galaxies twice as far away, travel twice as far through space meet/interact with twice as much matter, lose twice as much energy, undergo twice the reduction in frequency and hence have their wavelength increased by twice as much. Simple isn't it?

Now the room didn't erupt into riotous applause but it brought home the fact that velocity was not necessarily what was being measured. Velocity was a theory which may or may not be correct applied to the experimental observations. Hubble summoned it up by saying, '*it's redshift that we measure and from now on redshift is what we will call it.*' This is point when the theory of '*Tired Light*' was born. Zwicky published a theory on this in terms of some sort of 'gravitational drag,' but it never really caught on as the mechanism was not very convincing. The Appendix gives an overview of Zwicky's paper

Chapter 2

Retro predicting

Papers and thoughts on Tired Light have continued to crop up from this time on since it is such a simple explanation of these redshifts. However, until now, they have been slapped down like some naughty child who keeps stating the obvious - but let's continue with our story for the time being.

In 1940 A Canadian astronomer Andrew McKellar was the first to discover matter in interstellar space when he observed the characteristic spectra of organic cyanogen. He also determined the temperature at which these molecules existed and thus the temperature of intergalactic space as -271^0C or about 2 kelvin (have you heard of that temperature before? Yes? I bet most of you haven't heard of McKellar!).

The next part of our tale involves the theoretical Physicist George Gamow who was born in Odessa in what was then Russia. Gamow worked on the WW2 nuclear program, is well known for his work on alpha decay and, having spent many years ripping atoms apart to make nuclear bombs tried his hand at putting them together. Gamow was supervisor to Phd student Ralph Alpher and they came up with the idea of how the lightest particles that made up the Universe (principally Hydrogen, Helium and Lithium) could form from the primordial soup that came from the Big Bang. Working no doubt with what he believed was a 'long drink' at his side (which in Gamow's' terms was a tumbler full of whisky) the theory is said to predict the ratios, or *'relative abundances'* of the light elements which is roughly 74% Hydrogen, 24% Helium with the rest of matter making up the other 2% which is roughly what they are found to be. The paper was published on April Fools' Day 1948 and one claim to fame is that Gamow added his friend Hans Bethe (who had nothing to do with the paper at all!) so that the authors would follow the Greek alphabet – Alpher, Bethe, Gamow!

Predicting relative abundances of the light elements is put forward as one of the three pillars of the Big Bang Theory (Redshift is another and the CMB is the third which we will come to later) but the problem is that in order to predict the ratios, initial parameters have to be put in and there is a certain degree of flexibility in these! Worse still the theory cannot get all the ratios correct at the same time. Put in one set of initial conditions may well get the ratio of H:He correct but the ratios of H:Li and H:Li come out wrong – one needs a different set of parameters to get these ratios correct (but the H:He comes out wrong then). Whilst the Big bang Theory may be able to predict the ratios if one inputs the correct parameters at the start I would be more impressed if it could get them all correct at the same time!

Gamow then realised that if the Universe started in one big hot explosion it would cool down over time and so he set about calculating what the present day temperature should be and came up with a value of *'less than 5K.'* Completely independently of course??? Not really, because he knew of McKellar's work and the temperature of 2K before he started. We know this because Fred Hoyle (who we will visit in a minute) told him of this result and Hoyle refers to this in his books. Regardless of this, Gamow in the following years recalculated and recalculated producing higher and higher temperatures until he fixed on a temperature of 50K.

Now Hoyle and his followers didn't like the idea of the Universe starting in *'one Big Bang'* as he once said on a radio programme and this is how the Big bang Theory was christened! By a comment of derision by Hoyle.

Hoyle and his team believed in a *'Steady State Universe.'* This does not mean a static Universe. In the steady state Universe it is still expanding and redshifts are still taken to be recession velocities. However in the Big Bang, all the matter that exists to this day was created in that one moment in time whereas in a steady state Universe, as the galaxies move further apart *new* matter is created in the space between them so that the Universe remains *constant* and in a *steady state*.

The *acid test* that would differentiate between the two theories was the temperature of intergalactic space. In the Big bang Gamow was promoting his 50K whilst it was said that in a steady state Universe the temperature would be zero.

In the early 1960 Robert Dickie predicted the existence of Cosmic Microwave Background Radiation (CMBR) having allegedly forgotten that Gamow had done this sometime before.

Whilst Dickie was clambering about on the roof of the Physics Department at Princeton University in the nineteen sixties, two scientists, Penzias and Wilson had a problem. They were trying to study clouds of Hydrogen in intergalactic space by detecting the microwaves emitted from them but try as they may, a background 'hiss' prevented them from completing this study. Whichever direction they looked or whenever they recorded (summer, winter..) it made no difference. The horn detected the same background microwave noise. The horn was infested with pigeons and so they thought it could be radiation given off by the warm pigeon droppings so they captured the pigeons and sent them off by company mail to the opposite ends of the US. Unfortunately they were homing pigeons so they came back pretty quick. So the scientists caught them again and the pigeons gave up their lives in the interests of Science. Forlornly as it happened because the background hiss stubbornly persisted.

Stumped as to what to do next, they rang the local University Physics Department which just happened to be Princeton. Dicke (who had probably been called down from the roof to answer the phone) took the call and immediately knew what it was. It was microwave radiation at a temperature of 2.73K as predicted by Gamow many years earlier. This confirmed the Big Bang Theory and showed the steady state wrong. However, we must not be deceived into thinking it has proved the Big Bang correct, it shows that of the two expansion theories the Big Bang was more likely to be correct and the steady state more likely to be wrong. Tired Light was never in this fight and was just an interested bystander.

Chapter 3

Roll out the crystal spheres

All subject areas have a half-life of knowledge or so we are told and this is the average time taken for half the facts in that area to be shown to be wrong. One doesn't know which fact will be the next to be proven incorrect or when a particular fact will be shown to be incorrect - but one can predict fairly precisely how long it will take for half the number of facts in that subject will be proven wrong. Medicine is an area with a very short knowledge half-life and in certain areas of medicine when a student actually qualifies, less than half the facts that were thought to be a fact when he/she began their training are now known to be wrong when they complete it.

Half-lives in Physics are a little longer but with the Big Bang Theory fast approaching its centennial it must be in danger of facing the chop. Indeed the cracks have already begun to show.

For instance, in the Big Bang, lots of exotic particles would have been formed – so many that they should be all around us even now. Problem is we can't find them and we should be able to!

Then there is the horizon problem. In the Big Bang <u>space</u> went zooming off in all directions at speeds greater than the speed of light (which space can but matter cannot) carrying the matter with it. Now for the Intergalactic medium to be at a constant 2.73K it must all be in thermal equilibrium which means that all the matter must exchange radiation. This includes matter at opposite sides of the Universe. To do this, the radiation being exchanged must travel faster than the speed of light – and it cannot.

Then there is the flatness problem. The future of the Universe depends upon how much matter there is within it since gravity should slow the expansion down. If the density of the Universe is too low then it may well slow the Universe down but gravitational effects would not be sufficiently strong enough to stop it altogether. The Universe would continue to expand but at a slower and slower rate; becoming a very lonely and cold place indeed. This is known as an open Universe. If the density is too high then gravity would stop the expansion and the whole lot would come crashing down until it ended in one Big Crunch! This would be a closed Universe. However, if the density of the Universe is equal to a critical density then gravitational forces would bring the expansion to a halt and it would remain in this highly unstable condition where one flap of a Vulcan's ears would bring it all crashing down. This is known as a flat Universe.

The Universe is flat. Very flat indeed and that is the problem. Why is it that of all the densities this Universe could have had it chose to be the one almost exactly equal to the critical density?

Of course that is not the end of it by far. All UK GCSE and A Level Physics courses have this *'fate of the Universe'* included in their specifications – including those which were introduced in 2015. Students are required to learn and regurgitate it in order to pass. The surprising thing is that it is known to be a load of cobblers! At the beginning of the twenty first century, Supernovae were used to see just what is going on with the expansion and it was shown that none of the fates described in school Physics syllabuses were fates at all. Indeed what was found was that the Universe is accelerating faster now that it had before. In 2011 Reiss, Perlmutter and Schmidt of the High z supernovae team were awarded the Nobel Prize in Physics for their work on this. So one wonders why the school specifications have not been changed.

To explain these problems it was needed to bring in the biggest crystal sphere of them all – inflation. With inflation, a tiny tiny time after the Big Bang ($10^{-32}s$) an incredibly small small region of space started to expand at an unbelievably large rate. So large a rate that it is equivalent to something the size of a pea expanding to an object $2x10^{15}$ light year across. But it does brush all the problems under the carpet – an incredibly large carpet!

Since the original volume was so small there would be few exotic particles in it and that is why we cannot find them now. Since it was so small originally it would all be in thermal equilibrium and flat. Very flat. The problem is, inflation is a theory without substance. For inflation to work we need hypothetical '*scalar fields*' and no-one knows what they are. It is a theory, a fudge factor if you like to keep the Big Bang Theory's head above water. Some scientists reported that they had found evidence for inflation in 2014 with gravity waves but after trumpet blowing news announcements and modest claims of not wanting to talk about Nobel prizes just yet it appears that their glory and hopes had turned to dust – dust in our galaxy as probably the culprit. These dreams too were dusted under the carpet along with the problems.

So, to quote Sherlock Holmes. '*Once you eliminate the impossible, whatever remains, no matter how improbable, must be the truth.*' We have clearly eliminated the impossible with the Big Bang so let's look at whatever remains as the truth.

Tired Light.

Chapter 4

Assumptions and yet more assumptions

Generally speaking if experimental observations can be repeated and continually give the same results then they are correct. So it is not the results we should question but the interpretation of those results. We must also examine what assumptions have been made and revisit those too.

We cannot argue with redshift (symbol, z). We know that characteristic emission lines in the spectra of distant galaxies have a longer wavelength on arrival at the Earth as those same lines measured in the laboratory. This may seem a rather pedantic way of stating it so why not say, *'they have a longer wavelength on arrival then when they set off.'* If one does this then you are falling into a trap of an assumption. The lines correspond to differences in energy levels within an atom. When an electron falls down from a higher energy level E_2 to a lower energy level E_1, the difference in energy is given out as a photon of light. Now to find the frequency, f of the emitted photon we use the equation $E_2 - E_1 = hf$ and then to find the wavelength λ we use $c = f\lambda$ so we will have been guilty of falling for at least two assumptions.

We have assumed that the plank constant h and the speed of light c are not time dependent. We have only been measuring these two constants for the last hundred year or so and therefore how do we really know they are constants and independent of time? The galaxies we use for these redshifts are millions of light years away meaning that these photons that we are measuring at the present time set off millions of year ago. Could it be that the Planck constant millions of years ago was smaller than it is today making the frequency *then* higher than it is *now* and the wavelength *then* longer than it is *now*. Thus these photons will appear to have been redshifted to a longer wavelength – but they haven't – it was just the time dependency of the Planck constant making it appear thus. For this to work the Planck constant must increase linearly with time and thus the redshift – distance relationship will follow the Hubble law. There are workers who believe this to be the case and have alternative cosmologies based on this.

Then there is the speed of light which we know (?) to be a constant. We are assuming that the speed of light *then* was the same as it is *now*. Could it be that the speed of light *then* was faster than it is *now* and that it decreases linearly with time? That means that in the past the frequencies would match those today but the calculated wavelength of the original photon would be smaller. This too would also explain the Hubble law. Yes, you've guessed it, there are people out there with alternative cosmologies based on a variable speed of light.

I stated earlier *'guilty of at least two assumptions,'* have you spotted any others? Correct. We have assumed that the energy levels in atoms which were again only discovered around one hundred year ago have not changed in the past hundred million years. These are big assumptions and we should not take them lightly. Have you spotted any others? Well yes otherwise you would not be reading this book! The Big Bang Theory assumes that as these photons battle their way across the universe for hundreds of millions of years they did not lose even the tiniest bit of energy. The theory assumes that no energy is lost in travelling for all these hundreds of millions of year.

For me, I may be proved wrong but I can accept that the Planck constant h and the speed of light c are constants and also that the energy levels in atoms have remained unchanged. Why do I accept this to be true? Because their value would have to vary with time in a very regular way and mostly linearly which I find hard to accept. Why should they be time dependent in the first place and then why should it be such a regular relation? I would need this link to be explained before I could accept these proposals.

No, for me it is the assumption that the photons travel for hundreds of millions of years across the Universe which is laden with dust and plasma and not even lose the tiniest bit of energy that is wrong. In the Big Bang this is ignored totally and goes against all I know and have been taught. Some energy is always transferred to other forms whenever anything moves. These Big Bang photons are assumed to be

100% efficient in their motion through thick and thin! It cannot be right!

I believe that they do lose energy as they travel along. As they lose energy their frequency reduces. As their frequency reduces their wavelength increases. They have been redshifted! So what of the Hubble law that shows the redshift to be proportional to distance. I rejected variable Planck constant, speed of light and energy levels in atoms because I didn't believe any relationship could be so simple - so how does a loss in energy account for this? Simple.

Photons of light from a galaxy twice as far away, travel twice as far through intergalactic space, interact with twice as much matter, lose twice as much energy, their frequency decreases by twice as much and so their wavelength increases by twice as much.

This is the Hubble Law explained in term of photons losing energy as they travel. If we lose energy we become 'tired' and this is why these theories are called *'Tired Light'* theories.

Tired light theories have often been dispelled by mumbo jumbo or because up till now no Tired Light theory has been able to come up with a mechanism to explain it. Until now that is.

Welcome to New Tired Light!

Chapter 5

Travelling light

Despite main stream science refusing to accept Tired Light theories they persist because it is the simplest explanation for the increase in wavelength. Remember Occam's' razor? This tells us that if you have two competing theories then the simplest one is more likely to true. Here we have two explanations of these redshifts. One states that the whole Universe has to expand whilst the other says. *'Nah! It's just the light losing energy as it travels along.'* I know which is simplest.

The problem is that these redshifts have been studied by cosmologists and so they came up with a cosmological explanation. But redshifts are an optical phenomena and it is to optics that the cosmologists should have looked.

So does light travel?

Well before looking at how light travels through the universe perhaps we should start down here on Earth and look at how light travels through a transparent medium such as glass. The simplest model of an atom in glass, (and for all atoms to be honest) has a central nucleus with electrons orbiting around it. The photons can be thought of as consisting as short bursts of electromagnetic radiation – electric and magnetic fields that vary in strength sinusoidally - in phase with each other but acting perpendicular to one another. When a photon interacts with an electron in an atom the varying electric field causes the electron to oscillate at the same frequency as the photon. All the energy of the photon is transferred to the oscillating electron. There is a delay before the electron emits a *'new'* photon having the same frequency as the one absorbed. This process repeats itself as the photon travels through the glass.

The photons travel at the speed of light between atoms but due to the delays suffered at each interaction the *average speed* of light in glass is reduced. There is no loss in energy from the photon to the electron because they are fixed in the atoms which are fixed in the block of glass. The electrons do not recoil during the absorption/emission process and consequently the collision is perfectly elastic. Despite glass not being crystalline, having its atoms arranged randomly, the photons manage to travel in straight lines and emerge in the same order as they entered – glass is transparent.

In some cases the electrons hold onto the incoming photons' energy and this is where we get 'absorption lines.' Electrons in single isolated atoms can only have certain energies and these are known as *'energy levels'*. If the energy of the incoming photon corresponds to the difference between two energy levels then the photon is absorbed, as before, but this time it will not be re-emitted. The electron will remain in this excited state. This is known as *'resonance absorption,'* and produces the black absorption lines in the spectra from distant galaxies. It is these same characteristic lines that that are compared to laboratory standards that are used to measure redshift in distant galaxies.

So what of the transmission of light in intergalactic space?

Intergalactic space is not *'empty'* but full of plasma. Here the electrons have sufficient energy not to be associated with a particular nucleus but they can move from one to the next.

Overall the plasma is electrically neutral as there are equal numbers of electrons and protons (75% of the universe being mainly Hydrogen) but the electrons are not 'free' as there are long range electrical forces acting between the charged particles.

Electrons in plasma can and do oscillate. If an electron moves forwards it makes the space in front of it negative and leaves the space behind it positive. Restoring forces act on the electron to bring it back to its mean position and restore overall neutrality in the plasma. The electrons perform Simple Harmonic Motion (SHM). The protons being much more massive stay approximately where they are. It is the low mass electrons that have to *dance* around the high mass protons. If an electron can perform SHM then it can absorb and re-emit photons of light and so the light from distant galaxies travels through the transparent medium of intergalactic space by being constantly absorbed and re-emitted – just as they do in any transparent medium. Again there is a delay between the absorption and re-emission process but this time the electron is not fixed in an atom but recoils both on absorption and re-emission. The photon loses energy to the recoiling electron in the plasma of intergalactic space both on absorption and re-emission. The '*new*' photon emitted has less energy than the one absorbed. It has a lower frequency and a longer wavelength. It has been redshifted.

As the photons travel through space they will be absorbed and re-emitted by the electrons in the plasma. Each time the electron will recoil and some energy will be transferred to the recoiling electron. The energy of the photon reduces, the frequency reduces and the wavelength increases. It is redshifted. The farther the photon travels, the more electrons it interacts with and thus the greater the redshift.

Chapter 6

Testing times

Before getting into New Tired Light in detail maybe we should stay with the observations and some of the tests that have been carried out to differentiate between static and expanding models of the Universe. As soon as one mentions a static Universe on internet forums within a nano-second, up crops a message in reply, '*Tolman surface brightness test,*' or '*what about supernovae time dilation?*' so let's dispose of these objections from the start in case the reader is asking that same question.

Line counting.

One way of distinguishing between a static or expanding universe is to look back in time. In an expanding universe things should have been more squashed together in the past because the universe was smaller back then. That is, the average distance between objects in the past should be smaller than it is now. In a static universe i.e. one that is not expanding, the average distance between objects in the past should be the same as it is now since the Universe is the same size. We will look at two types of object, Hydrogen clouds and Supernovae Ia and see just how the average distance between them has changed (if at all) over time.

Hydrogen is the most common element within the Universe – there is no arguing with that and forms about 75% of all the atoms present. However they are not evenly distributed and tend to hang around in huge clouds which can be up to 11,000 light year from one end to the other. The clouds are distributed throughout the entire Universe with most of the Hydrogen atoms in them singular and not molecular (two Hydrogen atoms joined) and so are known a 'H1 clouds'. The atoms are electrically neutral consisting of a single proton as nucleus with a single electron orbiting around it. The electrons are most often found in their ground state with the electron orbiting in its lowest energy level.

Now we need a background light source – something very bright and very far away.

For this will use quasars (an acronym for Quasi Stellar or Star-Resembling). These are extremely small bright objects with many of them being far off towards the edges of the visible Universe. Now I know some readers will be tearing their hair out at this as scientists such as Halton Arp have disputed this for years and others, during and after his death, continue to do so to this day - but let's just stick with the mainstream view for now and have a chat about this at the end of the section. So light from these quasars can have been travelling across the Universe from the beginning of time and as it did so it will pass through H1 cloud after H1 cloud on its way to us.

Every time the light passes through a Hydrogen cloud some of the photons of light are absorbed by the electron in a Hydrogen atom. The electron leaps up to a higher energy level doesn't like being there and falls back down giving out another photon with energy equal to the difference in levels – but not necessarily in the same direction as before. It is not any photon of light from our distant quasar this happens to only 'special ones.' By 'special' I mean their energy must correspond exactly to the difference in energies between the ground state and the first excited state (or the next energy level up). Since the H1 clouds are so large virtually all photons of this particular frequency are absorbed and never to be seen again. This gives a black line in the spectrum because at this particular frequency there are now very few photons of this frequency left.

The remaining photons carry on until they meet the next Hydrogen cloud but the whole spectrum will be redshifted in the intervening space either by expansion or NTL. The black absorption line will move towards the red end of the spectrum along with all the other wavelengths but importantly, by the time the light reaches the next H1 cloud other photons will have been redshifted into that 'special' wavelength. The second H1 cloud will absorb all of these photons and so we will have two black absorption lines on our spectrum. On their journey to the third H1 cloud all photons and the two lines will shift towards the red end of the spectrum and a new group of photons will move into the 'special' wavelength. By this time the light reaches Earth it can have travelled through hundreds, if not thousands of H1 clouds and each one will have imprinted a black line on the spectrum. The spacing between consecutive black absorption lines indicates how far apart the H1 clouds are and so we have a ticker tape showing us the dynamic of the Universe – the biggest ticker tape on Earth! The 'special lines' are known as the Lyman alpha line as it occurs in the UV band of the e-m spectrum and the whole group of lines are known as the Lyman-α forest.

Now as any school child knows how to read a ticker tape. In a static Universe of constant size the H1 clouds should, on average, be equally spaced throughout time. In a Universe that is expanding, as the light travels across the Universe, time goes on. Initially on its journey the H1 clouds are tightly packed into a small Universe and so the lines should be close together. As the light nears the Earth the Universe will have become bigger during the time the light has been travelling and so later in its journey the H1 clouds should now be further apart and so the Lyman-α lines should be farther apart.

To put it simply, in a static Universe the lines should be evenly spaced whilst in an expanding one they should become farther and farther apart as the light approaches the Earth. There are some 'intrinsic redshifts' of course such as gravitational redshift but generally, cosmological Redshift is a measure of distance and hence time in any theory. To test the two theories what we can do is 'line count.' The redshift of the spectrum is divided up into divisions (called 'bins') of say 0.5 and the number of lines within this division are counted to give the number of lines per unit redshift (dN/dz). As the redshift increases we are looking farther and farther back in time and so in a static Universe as the redshift increases on average the number of lines per unit redshift (dN/dz) should remain the same. In an expanding Universe, as redshift increases, we are farther back in time when the Universe was smaller and the H1 clouds squashed together, the line count (dN/dz) should increase. Hence the Lyman-α forest will give us direct evidence of the dynamics of the Universe.

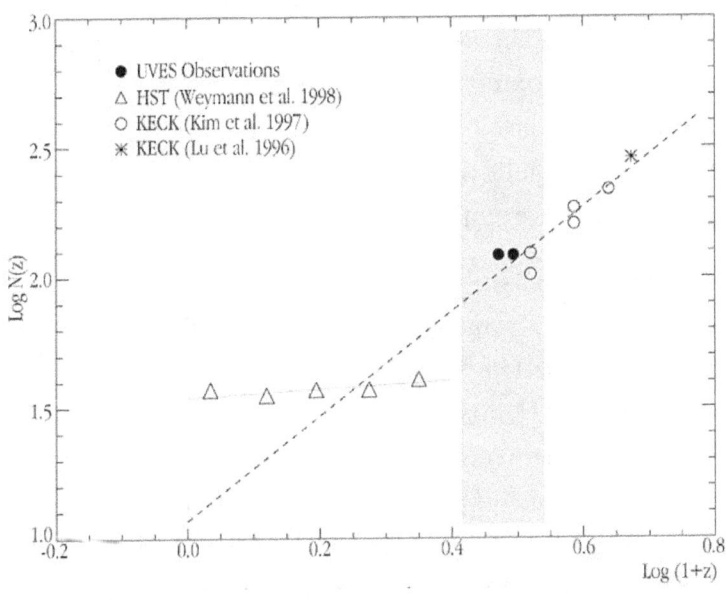

This image shows the number-density evolution of the Lyman alpha lines, observed with UVES in the quasars **HE22-28** and **QSO J2233-606**.

At first the data supported an expanding Universe with the line number density reducing with redshift i.e. as the redshift reduced, the clouds moved further apart. This part of the Universe had grown older and 'expanded' and so the

H1 clouds were farther apart on average. The data can be found at https://www.eso.org/public/images/eso0013g/ .

As said before, the Lyman-α lines are in the UV part of the e-m spectrum but these lines had been redshifted so much that when they reached Earth they were now in the visible region. To find out how the Universe was evolving in recent times we needed to look in the UV part of the spectrum as the lines would not have been shifted as much. When these results came in they showed a fairly constant line count up to a redshift of approximately 0.4. That is the Universe has been static for the last billion years or so! Taking these results at face value tells us that the Universe did expand in the past but stopped around a billion years ago.

This is interesting as if you remember the times before inflation (no, I don't mean the financial inflation) the possible 'fate' of the Universe depended upon the density? Too small and the Universe would continue to expand forever (an open Universe). Too high and gravitational forces would bring the expansion to a stop before bringing it crashing down into a big crunch (a closed Universe). However if the density is 'just right' and at a critical density then the Universe would expand, gravity would slow it down until the Universe came to a stop and then it would just stay there in the highly unstable but stationary form (a flat universe). It is known that the Universe is flat; very flat if not too flat! It is known that out of all the densities the Universe could be it just happens to be very close to the critical density. In fact as of 2013 the WMAP observatory showed the Universe is flat to within a 0.4% margin of error.

The true meaning of this is that the Universe is infinite but to kowtow to the Big Bang theory cosmologists say 'All's well. It's just that the Universe is much bigger than the bit we can see!' If that's not brushing the problem under the carpet I don't know what is. What is it they say, out of sight out of mind?

Not stopping there, as we saw before, a new theory was dreamt up known as inflation where a tiny part of the initial bang suddenly went berserk and expanded at an extraordinary rate. All well and good but there is still no evidence of this having occurred; nor is there any real mechanism for it to happen. As the Greeks used to say 'bring in another crystal sphere.'

Now this scenario is just what the H1 cloud data shows if we take it at face value. The critical density was 'just right.' The Universe expanded, gravity brought it to rest where it has stayed in this unstable state for the last billion year or so. New Tired Light is wholly responsible for the redshifts in this region and if you really do believe in Supernovae time dilation then these supernovae Ia are so far away that they are in the region where the Universe was still expanding! If this is the case then the Universe is much, much older than the 13.6 billion year estimate based on the Hubble constant. Even further out than that of the Bishop Usshers' who put the birth of the Universe at 4004BC (James Ussher was Archbishop of Armagh, Ireland who went through all the 'begats in the bible from Adam and came up with a day and time for the creation).

Some of you may be thinking that it's still a little improbable that the initial conditions were so close to the critical density. It goes worse if we accept the Universe did start in a Big Bang as the further and further we go back in time the closer and closer the density had to be to the critical one. Indeed, if this is the case then the density in the early Universe had to be within 1 part to 10^{62} of the critical density for this to happen. So why is this? It could be down to Darwin, survival of the fittest. There may have been lots of Bangs. Those with too high a density rapidly came to a crunch. Those Bangs with too low a density would expand forever with their component parts getting so far apart there is no sign left. Remember, the 13.6 billion year estimate is based on the Hubble constant which we now know has nothing to do with expansion but is down to Tired Light.

However, this is just one data set. Some workers using H1 Lyman-α line counting put the number count constant out to a redshift of 1.6 much further than those cited - here implying a very static Universe indeed.

Chapter 7

The Tolman Surface Brightness Test

The surface brightness test (SB) was put forwards in the 1930's as a way to discriminate between static and expanding universes. SB is the brightness per unit area of the galaxy i.e. the brightness of a standard sized 'piece' or area on the surface of the galaxy. One of the advantages of this method is that the SB relies only on whether the Universe is static or expanding and nothing else. The value of the Hubble constant, *'dark energy,' 'dark matter'* and any other mumbo jumbo make no difference. If the Universe is expanding the SB varies one way and if the Universe is static it will vary in another way. The SB will reduce as distance and redshift, z increase and so it should be proportional to $(1 + z)^{-n}$

In an expanding Universe n has the value of four. If the galaxy is moving away from us then fewer and fewer photons will arrive each second and this is responsible for a power of 'one'. Since the wavelength of the photons is larger due to redshift then the frequency will be lower and hence the energy carried by them is less. This contributes a further power of 'one'. In an expanding Universe the galaxy was closer to us when these photons were emitted and so the galaxy had a greater angular size. Again the factor involved is $(1 + z)$ and since it is an area we are interested in then we must square it. This contributes a further power of 'two.' Consequently, in an expanding Universe $n = 4$ and the SB will fall off rapidly as the distance and redshift increase.

In a static Universe, the only factor involved is the photons carrying less energy and so here $n = 1$. Consequently by measuring the SB of similar galaxies near and far to determine the value of n it should be possible to tell if the Universe is expanding or not.

Before looking at the results we need a word of caution here. In order to compare two galaxies they must be identical and clearly here is an assumption. How can we be sure that they are identical? Hubble introduced a classification system for galaxies but can we assume that two galaxies in the same classification have the same overall brightness? Then there is the problem of evolution. If one galaxy is further away then it takes longer for the light to reach us than that from the nearer one. Light from the more distant galaxy must have set off before that from the nearer one for them both to arrive at the same time. So the galaxies may be identical but the further galaxy was older than the nearer galaxy when the light was emitted. Consequently, we are not comparing like with like as the two galaxies are of different ages and any difference in SB could be due to an aging process or *'evolution'* of the galaxies.

But let's not bother with all that for now and look at the results. Does $n = 1$ or $n = 4$? The paper often quoted as '*proof*' that the Universe cannot be static using the SB test was published in 2001 and gave the results for the value of n for two bands or colours of light using filters. In the 'R band' (red) they found $n = 2.59 \pm 0.17$ whilst in the 'I band' (Infrared) they found $n = 3.37 \pm 0.13$ and stated that these values ruled out a static Universe but not an expanding one.

Now let's just think about this a minute. Firstly it does not say it has proven an expanding Universe - just that it has ruled out a static one. Secondly has it really ruled out a static Universe? In a static Universe $n = 1$ whilst in an expanding one $n = 4$. The average of these two values is 2.5 and isn't that what was found in the R band? All this result has done is split the difference? Thirdly, in the I band, 3.37 plus an uncertainty of 0.13 gives 3.5 which is still a long way from the $n = 4$ needed by an expanding Universe.

So these results may look convincing to a died in the wool expansionist to cynics like me they leave a lot to be desired. The Tolman test had not really been tested over a wide range of redshifts and so perhaps if this was done then the outcome might be more decisive. Eric Lerner, Renato Falomo and Riccardo Scarpa did just this and used disc galaxies drawn from the HUDF and Galex data sets.

The HUDF or Hubble Ultra-Deep Field is a set of data relating to approximately 10,000 galaxies gathered over a period of around five months as the Hubble telescope looked deeper and deeper into space. The area itself looks like an empty bit of night sky just Southwest of Orion with no foreground stars. Whilst the area is tiny – think of holding up a piece of graph paper at arms length then the 10,000 galaxies are continued within one small 1mm×1mm square.

GALEX or Galaxy Evolution Explorer is another orbiting space telescope recording galaxies in the Ultra-Violet across ten billion years of the history of the Universe. Both telescopes measured in the UV and are up in space as UV light is absorbed by our atmosphere.

Between the two sets of data we have a great range of redshifts from local to a redshift of 5 and so if there is something in the Tolman Surface Brightness test then it should show up here.

So, What did it show?

It showed:

'We conclude that available observations of galactic SB are consistent with a static Euclidean model of the Universe.'

That is the Tolman Surface Brightness test agrees with a static Universe.

Chapter 8

Alcock-Paczynski cosmological test

This is a nice little test as it does not need any of the mumbo-jumbo parameters such as dark matter or dark energy inputted to carry out the test. Evolution or '*aging*' of a galaxy does not matter either as the test purely depends upon the geometry of the Universe. All that is needed is to use anyone's theory to calculate a relationship between the redshift and the galaxy's' apparent diameter (how big it appears to be from earth). This is then compared to observation to see if that theory will work. Simple.

In 2014 Martin Lopez-Corredoira did just this and published the paper in the Astrophysical Journal. The author took six cosmological theories and used the test to try them out. The theories were: ΛCDM, Einstein-de Sitter (a theory using no dark matter), Friedman model (no dark energy), quasi-steady state cosmology (no big bang), a static universe model, and static universe with "tired light."

ΛCDM or 'Lambda Cold dark matter' is the standard mainstream model of the Universe with expansion. Lambda is the fiddle factor or cosmological constant. The static universe with 'Tired Light' Needs no explanation and we can forget about the other four as they are wrong!

What was shown was that only two of the six theories passed this test. They were the ΛCDM model and Tired Light. That is according to this test alone, the Universe is expanding as per mainstream or it is static with New Tired Light.

Chapter 9

Supernovae Time Dilation.

Until the turn of the century, there had been very little 'direct evidence' for an expanding Universe. It was just redshift, CMB and a claim that the Big Bang Theory could predict the ratios of the light elements (Hydrogen, Helium and Lithium). As I said earlier, I would have been more impressed if it could do it at the same time - mind you as certain assumptions have to be 'inputted' to predict these ratios and if it gets H-He ratio correct then the others are H-Li is way out and so on!

So Supernovae Time Dilation was the stuff of Nobel prizes. What was said was that these Supernovae all went with the same bang and could be used as a standard candle in cosmology. Standard candles are objects of the same or known Luminosity. Originally it was thought that they consisted of a stable white dwarf star with a companion close to it. The white dwarf star stole mass from its companion until it reached a certain mass (Chandrasekhar limit). Above this mass the star would erupt as a supernovae. Since the star built up to this mass they should all go with the same bang at the same mass and thus have the same Luminosity. We now know this is incorrect but no-body seems to mind (bring on another crystal sphere). Originally Astrophysicists measured their intensity over time as the supernova rose to full brightness and then slowly faded away but now they look at the spectra over the same period and measure the time it takes for certain spectral features (emission lines) to appear. The observational result is that the higher the redshift the longer these supernovae take to rise to full brightness and fall. This is interpreted – I will repeat that it is interpreted (not measured) as relativistic time dilation. Let me explain.

I am sat here tapping away at my computer keyboard. I have the time displayed both at the bottom right of my computer screen and the also by the clock across the room sitting on top of the television. They both say 10:30 pm (half an hour to pub closing). Now if I was to move the television a long way away so that the time taken for the light to reach me from the clock on the TV was two minutes. At 10:35 pm both clocks would give the same time, 10:35 pm and this is what I would see on the clock on my computer screen as it is right next to me. However, when I look at the clock on the TV, it would show 10:33pm because I am seeing it not as it is now but as it was two minutes ago. The two minutes being the time it took for the light carrying the information to reach me.

This is not time dilation. I just see the clock as running two minutes slow.

For time dilation the clock must be moving away from me very fast. Let's say that during every 5 minutes the television and clock move so far away from me that in this extra distance travelled the light takes an extra minute to reach me (and I only check the time every 5 minutes). So at 10:35 pm the T/V clock reads 10:32 pm since I am now seeing it as it was three minutes ago (1 minute plus the original two). At 10:40 when I look at my computer clock it says 10:40 pm but the T/V clock shows 10:36 pm as the light now takes four minutes to reach me. When I check the time at 10:45 the T/V clock shows 10:40 as the light now takes five minutes to reach me due to the extra distance the light has to travel. At 10:50 on my computer screen the T/V clock now reads to me 10:44pm as it goes further and further behind. I see it as the clock running slow but in fact the T/V clock is running perfectly – it is just that it takes longer and longer for the light to reach me. Now if you will excuse me I must go if I am to catch last orders.

Ahhh! I needed that!

Now the so called 'time dilation' is expressed as an 'aging rate' and it said that this aging rate is consistent with the factor $1/(1 + z)$ and whilst it didn't exactly agree entirely, it was said that there was enough 'time dilation' to rule out a static Universe and thus Zwicky's 'Tired Light' hypothesis. As always in cosmology, uncertainties in the measurements are huge.

But it is not as simple as this. What was measured and '*found*' was that the time for the supernovae to brighten then dim increased the farther way they were. Could there be other reasons for this to happen other than by relativity?

One possibility is pulse broadening by dispersion.

Dispersion is just like the spectrum we obtained at school when passing white light through a glass prism. Red light travels the fastest in glass so undergoes the least change in direction Violet travels the slowest and so deviates the most on being refracted. When a rectangular multi-colour pulse is sent down a fibre optic one gets 'pulse broadening' as some frequencies of the light travel faster than others. The pulse becomes broader and has a lower maximum as it travels along so that the area underneath it remains the same. The further it travels down the fibre the broader it becomes as the faster frequencies leave the slower ones further and further behind. Supernova events are multi-coloured and so one would expect a similar sort of thing. Originally they looked at the width of the light curve but more recently they look for the time it takes for certain frequencies (colours) to appear as spectral lines in the curves. Well, since all frequencies do not travel at the same speed through the intergalactic medium, dispersion will have an effect and could broaden the light curves. This technique is used in radio telescopes and is quantified by 'Dispersion Measure – DM.' Lower radio frequencies arrive at the Earth after the higher frequency signals even though they are from the same event and from the delay the intergalactic electron number density can be found. The more electrons there are in the path of the radio signal, the greater the delays.

Another possible explanation is the Malmquist bias. This is an effect in observational astronomy, first put forward in 1922 by Swedish astronomer Gunnar Malmquist, whereby astronomers show an inclination to looking at brighter objects rather than the dimmer ones. It just so happens that supernovae Ia explosions are not all the same. Some are brighter than others and the brighter the supernova the broader its light curve. In order to compare one supernova with another, firstly a template has to be fitted to standardise the curves. However, as we look farther and farther into the distance, the dimmer supernova are too faint to be seen and so we are looking at brighter and brighter supernova events and brighter and brighter supernova have broader and broader light curves! Could supernova light curve broadening be a Malmquist bias?

But then again, originally it was thought that supernovae Ia (SN Ia) were caused by a stable white dwarf star collecting mass from a nearby neighbour until it reached the Chandrasekhar limit when it went 'bang!' In 2015 it was reported that not all SN Ia's were produced in this way as some were caused by the collision of two white dwarf stars. The researchers said that *'not to worry, don't give your Nobel prizes back yet,'* whilst the supernovae started life in differing ways they all faded in a similar way and so could still be compared. However, it must be said that they are not all the same as was first thought when the time dilation theory was tested. Could supernova light curve broadening be because the original assumption about them being '*standard candles*' was wrong?

This author, me, in a paper published in 2016 decided to test the supernova data in another way. The same way as the mean H1 cloud separation was used to check if the Universe was expanding or not. There is now reliable data on 580

SNe Ia in the 'Union2.1' SN Ia Compilation Magnitude vs. Redshift Table at the supernova cosmology project and better still, it's on open access. I chose bins of redshift size 0.05 and went through the 580 supernovae placing them into their respective bin according to their redshift. It appears that this has not been done before (well I couldn't find it anywhere) so clearly it is time to have a look. Just as in the H1 cloud separation, in an expanding universe as we look back in time the universe was smaller then and so the supernovae events, on average, should be closer together in redshift as the redshift increases. In a static universe the supernova should, on average, be evenly distributed in redshift. I appreciate that this is just a simple test but sometimes simple tests are more compelling.

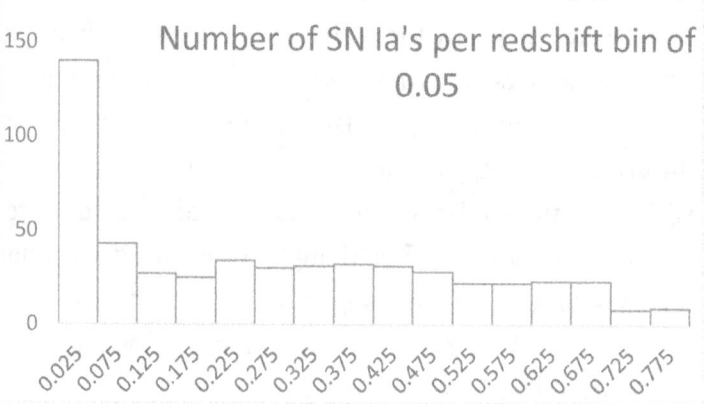

Notice firstly that there is a strikingly high number of SN Ia's between $z = 0 - 0.05$ compared to the much lower number distribution at a higher redshift. Note also that for redshifts above 0.05 the distribution is fairly flat and consistent with a static universe. It is in this region, the one where on average the Sn Ia's are equally spaced that the Supernovae said to be showing time dilation are to be found. How is it that on the one hand the SN Ia's showing time dilation and hence expansion at relativistic speeds can, on average, equally spaced? Problem!

Now let's just have a think as to why there are so many more SN Ia's close by than further out. Could it be that there are two types of SN Ia's - with one set not as bright as the other and so only 'local' ones are seen of the first type? If so, this would have an impact on the broadening of light curves and time dilation results since the local SN Ia's are used to set the templates on the basis that they are said to have negligible broadening. These templates are then used on the more distant ones to measure how stretched they are. In 2015 it was reported that there were two types of SN Ia which they labelled NUV-red and NUV-blue (NUV – Near Ultra Violet) and their luminosities vary in the UV section of the spectrum indicating that they may be caused by different events. Here we have a property of SN Ia that changes with distance. Importantly, in the nearby Universe almost 70% of SN Ia's are NUV-red and farther away the percentages fall such that only 10% of SN Ia's are now NUV-red. This would imply that a light curve template found from an average of nearby SN Ia's would not be valid when applied to the more distant ones ie the template used to calibrate the light curves is set using a nearby sample of SN Ia's consisting of 70% NUV-red and 30% NUV-blue and this is then used on distant SN Ia's which consist of 10% NUV-red and 90% NUV-blue. Could supernova light curve broadening be because of the gradual change in the sample from a nearby sample rich in NUV-red to a faraway sample rich in NUV-blue?

Chapter 10

Quasar Light Curve Non - Time Dilation!

Supernovae Ia are not the only heavenly bodies to have light curves. Quasars have light curves too and this is one of cosmologies best kept secrets! Why? Because they disagree with the supernovae results on time dilation and expansion – and we can't have that can we? M.R.S. Hawkins of the Royal Observatory in Edinburgh is one of several researchers who have looked for 'time dilation' in Quasar light curves. Using light curves from over 800 Quasars with timescales varying from 50 days to 28 years and small to large redshifts they looked for any evidence of time dilation and the conclusion was; *'The main result of the paper is that quasar light curves do not show the effects of time dilation.'*

So Supernovae light curves show time dilation but quasar light curves do not. Isn't this a little bit contradictory? If the stretching of the light curves is due to relativistic effects of the stretching of space then it should not matter what the source is. Either they both show time dilation or neither does if it is really due to the expansion of the intervening space.

So just why do the two results contradict each other? Again it could be that curve broadening is due to dispersion.

Supernova events are a multicolour pulse travelling through space and since different colours/frequencies travel at differing speeds the pulse is broadened. Quasar light curves are continuous so any dispersive effects would overlap and the only effect we would see would be at the start and end of the quasar itself – it would not show in the snapshot we have been given the privilege to see. It could be a Malmquist bias with the supernovae where as we look further and further into deep space we can only see brighter and brighter supernova and it is known that they take longer and longer to rise and fade. It could be that the supernovae time dilation result is just wrong because the sample used to set the template to calibrate the curves does not represent the sample of SN Ia's found in the distance. What do you think?

Name dropping here, I actually met Michael at the CCC2 conference (Crisis in Cosmology Conference number 2) at Port Angeles Washington State in 2009 where we both presented papers. We were all in awe with this guy from the Royal Observatory and someone asked, 'has anyone ever repeated your work?' to which he replied something like, 'not to my knowledge. The data is there and anyone can have it – they just haven't asked.' I wonder why not? Would it not be PC to prove the Big Bang wrong?

Whatever the case, it is clear that we should not rely on Supernovae 'curve broadening' as proof of an expanding universe until we can explain why Quasars do not exhibit the same 'curve broadening.' Both sources should exhibit the same phenomena if the stretching is due to the intervening space between source and observer. Either both should show curve broadening or neither. Until these two results can be reconciled then neither can be used as proof of anything. Expansion proponents will cherry pick the supernovae results whilst clearly proponents of a static Universe will cherry pick the Quasar results! The question remains 'why do expansionists ignore the Quasar results?

Chapter 11

Simple Harmonic Motion

As a result we can categorically tell people when they come up with these objections to a static Universe and Tired Light theories that they are out of date and cosmology has moved on to such an extent that these objections are no longer valid.

So let's return to New Tired Light. We saw earlier in NTL that the photons are absorbed and re-emitted by the electrons in the plasma of intergalactic space. Surprisingly a large number of people believe that a plasma is like a gas where the ions are free to move randomly as they like. They are certainly not *'free'* in the true sense of the word as there are long range electrostatic forces acting on the ions and whilst one may well have short range fluctuations in the charge of the plasma, overall these forces will keep a balance of neutrality.

Electrons having the smaller mass will move more quickly as temperature is a measure of the mean kinetic energy of the particles (protons and electrons since most of the plasma is made from ionised Hydrogen). A smaller mass means the speed must be greater to have the same KE. The electrons are more mobile and can and do perform Simple Harmonic

Motion (SHM) since, as an electron moves forwards the space in front of it has *'gained'* an electron and an overall negative charge whilst the space behind has *'lost'* an electron and now has an overall positive charge. The electron will experience a restoring force trying to bring it back to restore neutrality and so it will perform SHM.

Our photon coming along with its electric field varying vertically (say) will force the electron to oscillate vertically. The photon has been absorbed. However, because the forces between the ions in the plasma are weak the electron will recoil in the forwards direction and thus perform SHM along the line of sight due to the recoil impulse. This causes the electron to emit a secondary photon sideways and we will see later that this forms the CMB. The vertical oscillations of the electron where most of the energy is stored causes a *'new'* photon to be emitted in the forwards direction which is to all intents and purposes the original photon. On emission, the electron recoils backwards which again sets it into SHM in a direction parallel to the original path of the photon and a second secondary photon is emitted in a sideways direction until all the recoil energy is radiated.

However, the energy of our 'new' photon is less than the original photon absorbed because some of this energy has been *'lost'* to the recoiling electron. This is emitted as secondary radiation i.e. it forms the CMB.

Since the energy is less, the wavelength is longer. It has been redshifted.

We can and have worked out the increase in wavelength

suffered by the photons at each interaction. The momentum of the incoming photon, p is given by $p = h/\lambda$ and since this is transferred wholly to the recoiling electron we can calculate the velocity v of the recoiling electron. Using the formula $KE = mv^2/2$ we can calculate the energy transferred to the recoiling electron and thus *'lost'* to the photon. We double this to allow for recoil on absorption and re-emission and calculate the reduction in frequency by $E = hf$ and the new wavelength is found from $c = f\lambda$.

Perhaps surprisingly each photon undergoes the same increase in wavelength at each interaction regardless of its initial wavelength. This has the value h/mc or $2.43 \times 10^{-12} m$. This value is also known as the Compton constant but NTL must not be mistaken for Compton scatter as this has serious flaws when attempting to explain redshifts.

Let's just stop and think about this recoil interaction for a moment. A number of people confuse NTL with Compton scatter which is definitely not the case. Compton scatter is the absorption and re-emission of a photon by a totally *'free'* electron – one with absolutely no forces acting on it. If there are no forces acting on it then the energy cannot be stored by the electron between absorption and re-emission. In Compton Scatter, these must be simultaneous events – the photon is re-emitted immediately as it is absorbed. There is no transfer of energy to the recoiling electron in Compton scatter as the events (absorption and re-emission) take place at the same time. Where Compton scatter does provide a transfer of energy and hence an increase in wavelength is when the electron is scattered off to one side and the photon to the other. Whilst we get an increase in wavelength here the light is scattered by Compton. After several of these Compton scatter events as the light travels towards us, the image will be blurred. This is not the case. Images are not blurred by redshift. It cannot be Compton scatter that causes the redshift.

This is not a problem with NTL since the electrons here have the ability to store the original energy as vibrational energy as they perform SHM. This gives us a delay between absorption and re-emission and thus the electron recoils along the line of sight. The light travels in straight lines and thus there is no blurring of the image. Granted two low energy secondary photons are emitted to the side but these have low momentum and in any case since they go out in opposite direction their effect is slight if any.

Another question that NTL answers over its rivals is '*if redshift is due to the absorption and re-emission of photons by electrons in the medium, why is there no redshift as light travels through glass?*' In order to recoil, the electrons must be loosely held together. We want the electrons to have forces acting on them to allow them to perform SHM but for redshift these forces must also allow the electrons to recoil. In dense plasma the forces acting are much stronger and the electrons hardly recoil at all. There will be no redshift in dense plasma. In glass the electrons are held in atoms and the atoms are held in the block of glass. There is no redshift in glass since the mass of the glass is large and recoil negligible. It is only in the '*squidgy*' plasma of intergalactic space that the electrons recoil and redshift occurs.

As the photons of light travel across the Universe they are repeatedly absorbed and re-emitted by the electrons in the plasma and are redshifted at each interaction. Hubble's' Law is often confused by interpreting redshifts as velocities. From an experimental and observational point of view it tells us that photons of light from a galaxy twice as far away undergo twice the increase in wavelength than that from the nearer one.

In NTL this is explained as a photon of light from a galaxy twice as far away, travels twice as far through the intergalactic medium, interacts with twice as many electrons, has its wavelength increased twice as many times and thus experiences twice the overall increase in wavelength.

Chapter 12

Bumped into you by chance!

Now we know by how much the wavelength is increased at each interaction we now need to know how many interactions each photon makes on its journey through intergalactic space. That is, how many electrons does each photon bump into on its journey? To find this we need the probability of this happening and that is known as the 'collision cross-section' and has the symbol, σ. In simple terms, σ is a cross-sectional area surrounding the electron and if the photon comes within this area it will interact with the electron, be absorbed and re-emitted. If the photon passes outside this area it will not interact and whiz past as if nothing had happened – which it hasn't. We will see that the collision cross-section that concerns us in NTL depends upon the classical diameter of the electron (which is constant) and the wavelength of the incoming photon.

Collision cross-sections are known from experiments involving low energy X-rays and atoms. Here scientists are interested in the absorption of a photon by an electron and so they talk in terms of 'absorption coefficients.' As we said, collision cross-sections are a probability and so the probability of the photon being absorbed or transmitted consists of two parts. The next bit is a little tricky to explain but please bear with me as it is important. Important because New Tired Light uses actual published collision cross-sections and so we need to fully understand where they come from.

1) The probability of the photon being absorbed in the first place is known and is given by, $\sigma = 2r\lambda$. Here 'r' is the '*classical electron radius.*' This is not the actual radius of an electron but a dimension used in particle physics to represent the '*size*' of the electron in terms of particle interactions. The 'λ' is the wavelength of the photon. Whilst the 'size' of the electron is constant and the $2r$ represents its diameter the larger the wavelength of the photon the greater the chances of it bumping into an electron and being absorbed.

2) After this we have to consider whether the photon is retained or re-emitted. The Lawrence Berkeley National Laboratory Center for X-ray Optics and Advanced Light Source (http://xdb.lbl.gov/xdb.pdf section 1.7) gives the '*photo-absorption coefficient, f''*' and this is the probability of the photon being retained once its energy has been absorbed by the electron.

3) The probability of the energy of the photon being absorbed completely and not re-emitted is derived from the multiplication law of probability:

$$P(A \text{ AND } B) = P(A) \times P(B)$$

So the probability of our photon's energy being absorbed by the electron and retained is:

P(absorbed) = (probability of being absorbed in the first place)x(probability of being retained)

ie $\quad\quad\quad\quad$ P(absorbed) = $(2r\lambda) \times$ f'

Or $\quad\quad\quad\quad\quad\quad\quad\quad$ = $2f'r\lambda$

The values of f' are again known and published. For Hydrogen (one electron) f' varies between 0 (energy of incoming photon well away from difference in energy levels within the atom and so the photon is re-emitted) to 1 (where the energy of the incoming photon is exactly equal to the difference in energy levels within the atom and so the photon is not re-emitted). Helium has two electrons and so f' has values from 0 to 2 but since we are dealing with electrons in the plasma of intergalactic space then f' here has values between 0 and 1 since we only have single electrons. If our photon has just the right frequency it is absorbed, sets the electron into oscillation and because of resonance the whole plasma is set into oscillation. We do not get the energy back! The photon has been absorbed.

Now only two events are possible and these are mutually exclusive: either the photon is re-emitted or it is not. If f' is the probability of the photon being absorbed then let q be the probability of it being re-emitted.

The probability of re-emission will be $2qr\lambda$ and since:

$$f' + q = 1$$

When the frequency of the photon is well away from the resonance frequency $f' = 0$ and so $q = 1$. As we said at the beginning, the cross-section for re-emission of the photon and therefore transmission light is $\sigma = 2r\lambda$.

The next thing we need to know in order to calculate a value for the redshift is the 'mean free path, l.' This is the average distance a photon travels between interactions on its journey through space. Sometimes it will travel huge distances, sometimes not as far but on average, just how far will it travel between interactions?

The mean free path depends on two things:

i)   The collision cross-section σ

The larger the collision cross-section the greater the chance of the photon interacting with the electrons on its journey and thus the shorter the distance, on average, between interactions.

ii)   The '*electron number density, n.*'

This is the average number of electrons per cubic metre of space. The more electrons are put in the way of the photon on its journey then the greater its chance of being absorbed and re-emitted. However, we must remember that as n becomes larger, the electrons become closer and closer together, the forces between the ions in the plasma increase and the less the recoil becomes. Less recoil means less energy loss and therefore the redshift effect disappears in dense plasma.

Determining a value for n has posed a problem for alternative cosmologies since without knowing just how many electrons there are on average in intergalactic space then it was all a case of '*hand waving.*' Differing alternative theories required a range of electron number densities for them to work ranging from n equalled one hundredth of an electron per cubic metre to thousands! It was all 'if n = this' or 'if n = that' then it works. But just what was it? What is the true value in space? NTL has consistently said that $n = 0.5$ electrons per cubic metre of space was the value. It has been published in paper after paper in copyrighted journals since 2006 and so this claim can easily be checked.

Then in 2016 an event was recorded and two independent workers realised that this event could be used to determine the actual electron mean number density. They decided to calculate the value and test the various Tired Light Theories.

The hand waving was over!

Chapter 13

Dispersion Measure

Independent research is a big thing these days. The internet is a great source of information giving access to published papers often free of charge. If one has to pay to download a paper then one can often find a free version online with the arXiv e-print system. The SAO/NASA ADS database is a superb source for finding papers and all data and information is there online. These days all publicly funded research that does not involve National Security has to be available freely online. With all this anyone can perform their own research.

Independent researchers don't have to work by themselves either. Internet forums are a good way to air your views but another more popular way is by email. Workers find other people working on the same area and form a group where ideas and information are shared. Well maybe not all ideas as you want to publish them yourself. These groups get bigger and bigger as more and more people are CC'd in as time goes on. Many are just *'lurkers'* who just like to follow what is being shared but others are very active sharing news items to keep everyone up to date. I belong to two groups, one with over 40 members and another with over 60! Not that I actually joined these groups – my name just appeared on the list and the emails began. However, I recommend them not just for keeping you working but more of a self-support group. It's good to know that if it really is crazy to believe in a static Universe then there are plenty more besides you who think the same way.

It was through one of these groups that I first met FRB's or Fast Radio Bursts to give them their full title. These are high-energy radio pulses that last only a few milliseconds. Their sources are unknown but lie outside our own galaxy, the Milky Way and have a range of wavelengths.

Dispersion, you will have met in school as the splitting up of white light into its component colours ie how a prism splits white light up into a spectrum of colour and how a rainbow is formed. This happens because whilst in a vacuum, all frequencies and wavelengths travel at the speed of light $3x10^8 ms^{-1}$ but in glass, they all travel at different speeds. Red travels the fastest and is refracted the least, violet travels the slowest in glass and so is refracted the most. Hence a rainbow is formed.

Now dispersion isn't just restricted to light; it is a phenomena related to all the e-m spectrum from gamma rays to radio waves. An FRB goes off in a far away galaxy and a short, sharp burst of radio waves are sent out having a broad range of frequencies. Standard Physics tells us that as these radio waves travel through plasma the frequencies and wavelengths will all travel at differing speeds. Surprisingly, where it was the longer wavelength in light that travelled the fastest, in the radio it is the shorter wavelengths that travel the fastest. We will revisit this in a later chapter and explain the effect when we apply NTL to dispersion. For now, the pulse is dispersed since different frequencies suffer differing time delays between absorption and re-emission on interaction with the electrons. Whilst all frequencies set off together within a millisecond or two they arrive at different times. By measuring the time delay between two known frequencies arriving one can determine a quantity known as the '*Dispersion Measure, DM*' and if one knows the distance to the FRB one can determine the mean electron number density of the plasma the radio waves travelled through.

Great! Well no actually as we don't know what causes these FRB's. We also do not know exactly where they come from other than they must be from outside our galaxy as the dispersion is far greater than anything in the Milky Way. So how could we possibly know the distance in order to find the average electron number density? We can't do it. Can't that is until the April of 2015 when the Dispersion Measure of FRB150418 and the redshift of the host galaxy were measured for the first time. FRB's are named after the calendar date they were recorded as FRB YYMMDD so this one was recorded on the 18th April 2015. The paper was published in February of 2016 and this is when the emails started!

John Kierein I believe saw the published paper and emailed the reference around the group along with the realisation that this could be used to find the actual mean electron number density in intergalactic space. Another member of the group, Louis Marmet decided to calculate the value of n since here we had the Dispersion Measure and the redshift of the host galaxy and once you have the redshift one can calculate the distance in any cosmological theory. Apologies if I have mixed up their contributions but John and Louis have been long time followers of alternative cosmologies and basically, this bit is down to them. As a check, Louis invited other members of the group to work blindly and calculate their own values so we could compare them later and ensure we had it right.

Well, you guessed it. The value of n from this FRB result was $n = 0.5 m^{-3}$ exactly as predicted by NTL all along. This was a game changer for NTL. In science we have always said that a true test of a theory is if it can predict something that is later proven to be correct then it is a very good sign indeed. With the Big Bang and Steady State it was the CMB. Big Bang predicted it. Steady state did not. So when the CMB was found the Big Bang became the accepted theory whilst the Steady State was banished to the dark corners of the Universe. So what of NTL? Surely this must now be THE Theory since its published prediction of over ten year ago has now been shown to be true.

We will return to Dispersion Measure again since in NTL both DM and NTL are caused by photon electron interactions then one would expect a link between the two – and there is. To be honest, I hadn't heard of Dispersion Measure before the emails between John and Louis and the group and here is a whole new area to apply NTL. As I said earlier, it's OK to be independent but don't be a lone worker.

We are now in a position to predict a value for the Hubble

constant so why not try it and see!

Chapter 14

The Hubble Constant, H

We are almost ready to jump in and start applying New Tired Light to not only explaining cosmological effects but by calculating quantitative values for redshifts, the Hubble constant etc.

To recap on our starting point in New Tired Light, the photon from a distant galaxy comes along, is absorbed by an electron in intergalactic space and the electron is set into Simple Harmonic Motion – it oscillates. Most of the energy of the incoming photon has been transferred to vibrational energy of the electron in the plasma. Some of the energy is transferred to the recoiling electron and as the electron recoils it is set into oscillation along the line of sight in addition to the transverse oscillations due to the photons' oscillating electric field.

A low energy photon is emitted to one side due to the oscillations along the line of sight caused by the recoil energy transferred and this forms the CMB. The transverse oscillations of the electron cause a new photon to be emitted in the forward direction but again the electron recoils causing a second low energy photon to be given off to the other side which also forms part of the CMB.

The photon continues in a straight line as the recoil takes place along the line of sight and so no blurring of the image occurs.

On each photon-electron interaction the wavelength is increased by an amount 2.43×10^{-12}m regardless of the wavelength of the incoming photon.

One must remember that the treatment here is a classical or non-relativistic one and breaks down when wavelengths of the incoming photon reach X-ray or gamma. At these wavelengths, the velocity of the recoiling electron would exceed the speed of light! Consequently we would have to rework the theory for photons of wavelength 10^{-10}m or smaller as clearly these could not be redshifted by the amount shown above.

Now lets take a look at where we want to get to – the redshift, z and the Hubble constant, H. I appreciate that the symbol for the Hubble constant is H_0 as in the Big Bang Theory the constant isn't constant but changes over time due to the effects of gravity and so on but we are above that here!

The wavelength of light of a characteristic line in the spectrum from the distant galaxy has a longer wavelength than that same characteristic line as measured in the laboratory on Earth. If the wavelength of this characteristic line in the laboratory is λ and the wavelength of this same characteristic line in the spectra from the distant galaxy is

λ', then the increase in wavelength, $\Delta\lambda$ is given by: $\Delta\lambda = \lambda' - \lambda$. The redshift, z is given by: $z = \Delta\lambda/\lambda$ and it is important to note that for a particular galaxy, the redshift z has the same value for all wavelengths. That is, the longer the wavelength, the greater the shift in wavelength such that the ratio $z = \Delta\lambda/\lambda$ remains constant.

Now at the beginning and for want of something else to call them (and for no other reason) Hubble converted these redshifts to Doppler velocities. As we know he reverted back to redshift when Zwicky pointed out that there were other ways of explaining the redshifts other than by the Doppler Effect. In the Doppler Effect, the *'velocity,'* v is given by $v = cz$. Having assigned a *'velocity'* to the redshifts (is it any wonder the Big Bangers lost the plot?) he then plotted the Doppler velocity, v against the distance to the galaxy, d and obtained a straight line through the origin showing that $v = kd$. Where k is the constant of proportionality or gradient of the graph. Hubble being humble didn't name the constant after himself – that was done by other scientists later in honour of the great man to give: $v = Hd$.

If one looks at the data Hubble had it is anything but a straight line through the origin – more like a scatter diagram with a positive correlation. So much so that he left it to a friend to present his findings and had to be persuaded for even that to happen.

The Hubble constant has a place high upon the pedestals of physical constants since, in the Big Bang Theory, it leads directly to the age of the Universe. In the BB all matter was created at the same point of creation and all the bits and pieces came whizzing out at differing speeds, v and differing directions. Since all the bits and pieces have been travelling for the same length of time, t the distance travelled, d in this time can be found from speed equals distance over time, or $v = d/t$. Since $v = Hd$ we can equate the two for v and rearrange to give the age of the Universe as $t = 1/H$. The latest value of H from the Hubble space telescope using direct measurements (ie not contaminated by BB theory) is $73.00 \pm 1.75 \, km/s \, per \, Mpc$ giving an age of the Universe of around 13.8 billion years old.

Have you ever stopped to think about the units of the Hubble constant? I mean $km/s \, per \, Mpc$? The 'km/s' is easy enough – that is just a velocity but '$per \, Mpc$'? well, Mpc is a distance too so why not use the same unit of distance twice ie $km/s \, per \, km$? Or even $Mpc/s \, per \, Mpc$? Doesn't make sense does it? The reason is to instill this notion of expansion into everyone from the start. A galaxy 1Mpc away from us has a recessional velocity of $72 \, km/s$. A galaxy 2 Mpc away from us has a recessional velocity of $144 \, km/s$ and so on. But when we use the same unit for distance consistently, $km/s \, per \, km$ or $Mpc/s \, per \, Mpc$ we see that the distance cancels. The true unit of the Hubble constant is s^{-1}. In SI units, $H = 2.4 \times 10^{-18} s^{-1}$ and this is the number we need to calculate from our New Tired Light Theory.

In NTL the Hubble constant is given by: $2nhr/m$ where n is the average electron number density shown earlier from

Distance Measure as $0.5\ m^{-3}$, h is the planck constant $6.63 \times 10^{-34}\ kg/s$, r is the classical electron radius $2.82 \times 10^{-15}\ metre$ and $'m'$ is the electron rest mass $9.1 \times 10^{-31} kg$. Substituting into our equation for the Hubble constant gives: $H = 2.1 \times 10^{-18}\ s^{-1}$ – a difference of just 12% on the present accepted value.

Don't believe me? Well lets do it another way. A Hubble constant of $72\ km/s\ per Mpc$ means a galaxy $1\ Mpc$ away has a Doppler velocity assigned to it of $72\ km/s$. So let's see if we can get there from first principles.

A typical photon of light has a wavelength of $5 \times 10^{-7} m$ and so its collision cross-section will be $\sigma = 2r\lambda = 2.82 \times 10^{-21} m^2$. The mean free path, l for this photon as it travels through the intergalactic medium is $l = (n\sigma)^{-1}$ and we alresdy know the mean electron density, n from the Dispersion Measure and it has a value $n = 0.5 m^{-3}$ and so the mean free path is $l = 7.1 \times 10^{20} m$. $1\ Mpc$ is $3.1 \times 10^{22} m$ and so in travelling $1\ Mpc$ our photon makes, on average, 43.7 interactions. As we saw before the increase in wavelength at each interaction due to recoil is $2.43 \times 10^{-12} m$. The total increase in wavelength for our photon as it travels $1\ Mpc$ is $\Delta\lambda = 43.7 \times 2.43 \times 10^{-12} = 1.1 \times 10^{-10} m$.

The redshift z is given by $z = \Delta\lambda/\lambda = 0.00022$. To compare this with a Doppler velocity we must use $v = cz = 3 \times 10^8 \times 0.00022$ or $v = 66,000 ms^{-1}$ or $v = 66\ km/s$.

That is NTL, from first principles gives a value for the Hubble constant of $66\ km/s\ per\ Mpc$ which compares well with the accepted value of $72\ km/s\ per\ Mpc$ and is certainly within the uncertainties in this measurement.

Impressed? You should be it took me twenty years to get to this point!

Chapter 15

Corbor Galaxy Cluster

In the last chapter we calculated the Hubble constant and were pretty much spot on. The value predicted by NTL was 64 $km/s\ per\ Mpc$ - a value you will find as a '*true*' value dotted across the internet as determined by various means. However, we used a formula derived algebraically to do this which is fine in a published paper but I don't want this to become a math book. That said there is no point in writing a book on '*Tired Light*' if you don't show what it can do. As a compromise here we will work from first principles and for those of you who want the math algebraically I will refer you to the published papers which can be downloaded free of charge (see end of book).

In order to derive a redshift and compare it to that from observation we first need to select a galaxy and the one I am going to select is the Corona Borealis cluster A2065 or CorBor for short. This is the richest cluster in the Corona Borealis supercluster and is quite famous. My reason for selecting this cluster of galaxies is that it is one of several used by Humason and Hubble in the 1930's to show that the Universe is expanding. It seems a good idea then to use the same cluster to demonstrate that the Universe is not expanding and the results obtained by Humason and Hubble can be explained by New Tired Light.

CorBor or A2065 is listed in the '*Abell catalogue of rich clusters of galaxies.*' This catalogue was compiled by Abell initially as part of his PhD thesis and contains over 4000 galaxy clusters with redshifts $z \leq 0.2$ and so forms a collection of galaxies to challenge amateur astronomers as they are not too difficult to find on a cold dark night. To be a cluster there must be more than fifty galaxies within a small defined angle and the more the galaxies within this defined angle the '*richer*' the cluster. A2065 has a redshift of $z = 0.0714$ and is a distance of 960×10^6 light year from us.

We will select the wavelength of one characteristic line in the spectrum from A2065 so how about the '*K Line*' of ionised calcium $\lambda = 3.964 \times 10^{-7} m$. We could have chosen any of the Characteristic lines but I thought that since Calcium gives you strong teeth we should be able to get our teeth into the problem!

The first thing we need to know is the collision cross-section, σ. In NTL this is given by $\sigma = 2r\lambda$ where r is the classical electron radius and equal to 2.82×10^{-15} m. This means our collision cross-section for this Calcium line is $\sigma = 2.23 \times 10^{-21} m^2$. As we said before, σ represents an area and we can think of it as if the photon passes within this cross-sectional area around the electron they will interact.

The next thing we must find is the mean free path, l. This is the average distance a photon travels between interactions. It depends upon the collision cross-section and the average number of electrons per cubic metre of space. The greater the collision cross-section or wavelength then the shorter the average distance between interactions. The mean free path comes from standard statistics and is given by the formula $l = (n\sigma)^{-1}$ and the mean electron number density is now known from Dispersion Measure and has the value $0.5\ m^{-3}$. Substituting tells us that the mean free path, $l = 8.96 \times 10^{20} m$ or $94{,}700\ ly$. That is our photon is absorbed and re-emitted on average every 94,700 years on its journey to us. Not very often is it!

CorBor galaxy cluster is 960Mly away and so our photon will make $960 \times 10^6/94{,}700 = 10{,}140$ interactions with electrons on its journey to us through intergalactic space.

Each time the photon is absorbed and re-emitted by an electron energy is transferred to the recoiling electron, its frequency reduces and its wavelength increases by $2.43 \times 10^{-12} m$. Since the photon makes 94,700 interactions on its way the total increase in wavelength, $\Delta\lambda$ is $\Delta\lambda = 10{,}140 \times 2.43 \times 10^{-12} = 2.46 \times 10^{-8} m$.

The redshift, z is the ratio of increase in wavelength to original wavelength or $z = \Delta\lambda/\lambda$. Consequently the redshift is $z = 2.46 \times 10^{-8}/3.964 \times 10^{-7}$ or $z = 0.062$. That is our prediction of redshift from first principles and using standard Physics is $z = 0.062$. If you remember, the observed value of the redshift is $z = 0.071$ – a difference of around 10% and well within the uncertainties of the observations.

If you remember, we chose this galaxy A 2065 because Humason and Hubble also used it in their observations. The overall conclusion was that the Universe was expanding. Here we see that a far simpler theory not only explains the effect but derives the actual value.

Why did the cosmologists have to invoke the whole Universe expanding in order to explain this result? Because they were cosmologists looking for a cosmological answer when they should have been looking to optics for an optical answer. These redshifts are caused by light interacting with the medium it is travelling through.

Chapter 16

The Cosmic Microwave Background Radiation (CMBR)

We all know the story behind the CMBR (or CMB as it is sometimes known). I gave a brief overview of the history behind the BB and the CMB at the start. I had tried to avoid it altogether and assume everyone reading this book had read the history three million times over (if not more) but it didn't really work so I did a brief synopsis at the start. Apologies to those who have read it till ad nauseum!

Anyway we need to address it here as it has become one of the *'pillars'* of the BB. It has to be said that this microwave radiation that exists all around us may have nothing to do with the BB at all – it may be something completely separate. However, Gamow and others, knowing it already existed at about 3K since it had been discovered and published in 1941, continually changed their predicted temperature up and down between 5K and finally 50K. When it was *'discovered'* in the 1960,s by Penzias and Wilson at 2.73K it was said to confirm the BB since it had been predicted!

Of course, anyone with a television set had already discovered it as the CMBR forms part of the white noise that one gets on terrestrial television when it is off tune.

Maybe every couch potato in the World should have received the Nobel Prize too along with Penzias and Wilson as they had not idea what it was either.

Let's put sarcasm to one side for a moment. This radiation exists, has a black body distribution of wavelengths and peaks at a wavelength of 2.1mm. This corresponds to the radiation emitted from a black body at a temperature of 2.73K. It is the same in any direction and is independent of time of year.

In the BB it is the radiation left over from the '*BANG*' that has been stretched and is now in the microwave region of the e-m spectrum. The hot Big Bang has cooled down so much that its' temperature is down to 2.73K.

In NTL, this CMBR is the secondary radiation emitted by the recoiling electron when it absorbs or re-emits a photon. In NTL, the microwave radiation is local and comes from plasma clouds surrounding our galaxy. The '*clumpiness*' is not some '*quantum mechanical fluctuations*' as the BB would have us believe, but due to separate plasma clouds surrounding our galaxy. Plasma clouds emit Black Body radiation so that point is settled straight away. The BB interpretation runs into problems with the Black Body Radiation which we will look into at the end of the book – I want to keep on track as yet!

It's an easy matter to calculate the wavelength of this secondary radiation – we just need to calculate the energy transferred to the recoiling electron. The momentum, p of the photon is given by $p = h/\lambda$ and once absorbed this is transferred to the recoiling electron, $p = mv$. Equating these for conservation of momentum and we have $h/\lambda = mv$ which allows us to find the recoil velocity of the electron. Once v is known we can calculate the kinetic energy transferred from $KE = mv^2/2$, the frequency of the emitted photon is found from $E = hf$ and from this the wavelength λ can be found.

We will not work out the values here as I want to finish off our 'first principles' calculations with a spectacular display of the New Tired Light.

So Let's cream it!!

Chapter 17

Creaming it!

The discovery of the Dispersion Measure for a FRB and the redshift of its host galaxy was a game changer for NTL as now we had the actual mean electron number density in IG space. This put an end to guess work and '*if this*' or '*if that*' we can now tell it as it is. Now we know how to calculate the redshift and the CMBR lets pick another galaxy and go for QED!

We are going to go back to first principles for the last time and look at the galaxy Messier 82 or as Hubble and Humason knew it, NGC 3034. The galaxy is also known as '*the cigar galaxy.*' It's quite an eventful galaxy lying in the constellation Ursa Major and is five times as bright as the whole Milky Way. It's centre is one hundred times more luminous than the core of the Milky Way, contains the brightest pulsar yet known and had its own Supernovae Ia event in 2014. M82 was discovered in December 1774 by Johann Elert Bode – the same Bode as is in Bodes' law (That the orbits of the planets in the solar system form a mathematical series). However, interesting as these facts are this is not the reason for selecting it for examination here. The reason, as before, is that it is one of the Galaxies

Humason and Hubble observed to produce the famous Hubble diagram, published in the ApJ in 1931. The same diagram used to show expansion. I don't blame Humason and Hubble at all. They were great men who expertly gathered the data. It was others who interpreted the data wrongly. Hubble changed his own mind later and decided that these redshifts did not represent velocities or expansion.

M82 has a redshift of $z = 0.00068$ and is at a distance of 11.42 *Mly* or $1.08x10^{23} m$ away from us. Lets consider a photon somewhere in the middle of the visible spectrum with a wavelength $\lambda = 5x10^{-7} m$.

Collision cross-section $\sigma = 2r\lambda = 2.82x10^{-21} m^2$.

Mean free path $l = (n\sigma)^{-1} = 7.1x10^{20} m$

In travelling a distance of $1.08x10^{23} m$ our photon will make 152 interactions with electrons in the IG medium. That's about one interaction every 79,000 years! On each interaction the wavelength is increased by $2.43x10^{-12} m$.

Total increase in wavelength, $\Delta\lambda = 152x2.43x10^{-12} = 3.7x10^{-10} m$.

Redshift $z = \Delta\lambda/\lambda = 3.7x10^{-10}/5x10^{-7} = 0.00074$

This compares well with the observed redshift of $z = 0.00068$ and represents a percentage difference of only 9%.

Now lets calculate the wavelength of the secondary radiation emitted by the recoiling electron. In NTL these form the CMBR and so they should be in the microwave region.

Momentum of incoming photon:

$$p = h/\lambda = 1.33 \times 10^{-27} Ns$$

By conservation of momentum $p = h/\lambda = mv$ giving a recoil velocity of $v = 1460 \, m/s$.

The Kinetic energy transferred to the recoiling electron $KE = mv^2/2 = 9.6 \times 10^{-25} J$.

This energy is emitted as a secondary photon and since $E = hf$ the frequency of this secondary photon is:

$$f = 1.4 \times 10^9 Hz.$$

The wavelength is $\lambda = c/f = 21 cm$. We see that this is in the microwave region. However with the CMBR the peak of the radiation occurs at 2.1mm and so these secondary photons must be formed by original photons in the UV.

The redshift z is the ratio $z = \Delta\lambda/\lambda$ and has the same ratio regardless of the initial wavelength. This is from observation and so any theory must comply with this. So far we have only shown the NTL theory works for one wavelength and so for completeness lets repeat the calculation for a UV photon of wavelength $\lambda = 5 \times 10^{-8} m$. (BTW. I am just going to copy and paste here changing the powers of ten so feel free to skim this next bit – BUT take great care to take note of the conclusions!)

Collision cross-section $\sigma = 2r\lambda = 2.82 \times 10^{-22} \, m^2$.

Mean free path $l = (n\sigma)^{-1} = 7.1 \times 10^{21} m$

In travelling a distance of $1.08 \times 10^{23} m$ our photon will make 15 interactions with electrons in the IG medium. That's about one interaction every 790,000 years! On each interaction the wavelength is increased by $2.43 \times 10^{-12} m$.

Total increase in wavelength, $\Delta \lambda = 15 \times 2.43 \times 10^{-12} = 3.7 \times 10^{-11} m$.

Redshift $z = \Delta \lambda / \lambda = 3.7 \times 10^{-11} / 5 \times 10^{-8} = 0.00074$

Which is exactly the same as before showing that in NTL redshift is independent of wavelength for a particular source. The reason is that the wavelength was less and so the collision cross-section was less. That made the mean free path longer and so the total increase in wavelength was less. The initial wavelength is less and the increase in wavelength is less such that the ratio $\Delta \lambda / \lambda$ remains the same.

Now lets calculate the wavelength of the secondary radiation emitted by the recoiling electron from a UV photon. In NTL these form the CMBR and so they should be in the microwave region.

Momentum of incoming photon:

$$p = h/\lambda = 1.33 \times 10^{-26} Ns$$

By conservation of momentum $p = h/\lambda = mv$ giving a recoil velocity of $v = 14600 \, m/s$.

The Kinetic energy transferred to the recoiling electron $KE = mv^2/2 = 9.6 \times 10^{-23} J$.

This energy is emitted as a secondary photon and since $E = hf$ the frequency of this secondary photon is:

$$f = 1.4x10^{11} Hz.$$

The wavelength is $\lambda = c/f = 0.21 cm$ or $2.1 mm$. These photons are exactly at the peak of the CMBR curve. The CMBR curve falls off beyond the UV as once we approach X-ray the interaction becomes relativistic and the classical treatment here does not apply.

We can finish off our *'first principles'* calculations by explaining just why there are so many CMBR photons – and there are a lot! They tend to be the most common in the Universe.

Consider our UV photon $\lambda = 5x10^{-8} m$ undergoing a redshift of $z = 0.1$. Since $z = \Delta\lambda/\lambda$ the shift in wavelength is $\Delta\lambda = 5x10^{-9} m$. At each photon-electron interaction the shift in wavelength is $2.43x10^{-12} m$ and so the number of interactions to give us a redshift of $z = 0.1$ is 2060. Since there are two secondary photons emitted at each interaction (one of absorption the other on re-emission) that gives a total of 4120 CMBR photons when our single UV photon undergoes a redshift of just 0.1. Is it any wonder that there are so many CMBR photons in the Universe?

Chapter 18

The Exponential function

When we worked out the predicted redshift for the CorBor cluster our answer was a little less than the observed value. It was still pretty good but could we achieve a closer prediction? The answer is yes and the reason is that we assumed the collision cross-section σ was a constant for the whole journey and this is not the case. The assumption does not have a great effect when the galaxy is close and all the galaxies in the Abell catalogue are relatively close but later as we go farther and farther out it will make a difference.

Our photon sets out with an initial wavelength λ but every time it suffers an interaction the wavelength will increase by $2.43x10^{-12}m$. After the first interaction its wavelength will increase as it has been redshifted and is now $λ + 2.43x10^{-12}$m. The wavelength has increased and so too must its collision cross-section which means the mean free path is also less as there is now a greater probability of it bumping into an electron and interacting. After the second interaction the wavelength is now even greater and equal to $λ + 2.43x10^{-12} + 2.43x10^{-12}$m and so its cross-section is greater still and travels even less between interactions.

The photon travels like a stone skimming across the water. The first skip is long, the second shorter, the third shorter still as our photon undergoes shorter and shorter distances between interactions. The mean free path depends upon the wavelength and as the wavelength increase the mean free path reduces. Whenever the rate at which something changes depends upon the initial value (wavelength here) we get an exponential function and the same is true with NTL.

To find the number of interactions each photon makes on its journey from the galaxy to us we have to add up all the separate mean free paths taking into account that they are becoming shorter and shorter as the photon travels. Mathematically this is the sum of an arithmetic progression and this is made easier by means of a mathematical trick. If the differences between each term are small and the number of terms large then the sum approximates to an integral. For a true algebraic expression of our Hubble law we must include the collision cross-section increasing as the photon travels and this gives us:

$$z = \exp(2nhrd/mc) - 1$$

Now for galaxies a small distance away we can use the standard approximation $\exp(x) = 1 + x$ and so for nearby galaxies our Hubble equation becomes;

$$z = 2nhrd/mc$$

But Hubble's Law is in terms of these dreaded Doppler velocities and has the form $v = Hd$ where $v = cz$ and so substituting for z gives us $v = 2nhrd/m$. Comparing the two and we have:

$$v = Hd$$

$$H = 2nhr/m$$

As we saw when we predicted the redshift of galaxies used by Hubble to determine the law it works very well with nearby galaxies but we must remember the true formula for Hubble's Law is:

$$z = \exp(Hd/c) - 1$$

Whish is precisely the formula predicted by Zwicky in his Tired Light paper! But this time we have a mechanism to back it up mathematically.

Chapter 19

Dispersion measure DM reloaded

We met DM earlier and this is how we were able to find the mean electron number density in IG space and confirm predictions made by NTL over ten years earlier. But we can take it further than that and this is an exciting development for alternative cosmologies as it allows us to test our theories on a different area of cosmology – one where in the main stream BB there should not be a relationship at all. And there is!

In school physics, Dispersion is the splitting up of white light into its component colours and is caused by different wavelengths (colours) travelling at different speeds in a medium. In a vacuum they all travel at the same speed but this is not so in a medium. The reason is that the photons are absorbed and re-emitted by the electrons in the atoms of the medium. In IG space it is the electrons in the plasma that are responsible for the delays. Whilst the photons travel at the speed of light between atoms/electrons when one takes into account the delays suffered at each interaction the average speed is reduced. Different frequencies suffer a different time delay and so the average speeds differ. Hence they are dispersed.

Dispersion isn't just about producing a rainbow by passing white light through a prism but affects all e-m radiation. A square pulse of white light passed down a fibre optic ends up broadened with red light arriving first and violet last due to dispersion and their differing speeds. We have already considered this as a possible explanation of supernovae Ia curve broadening and the so called *'time dilation.'* Gamma rays too show dispersion as the *'magic'* project detected from gamma ray bursts.

It is in radio astronomy that the excitement begins as here too different frequencies travel at different speeds through the plasma of IG space. Importantly, there is a known relationship between the time delay between two frequencies arriving and the mean electron number density experienced on their journey. That is, a brief radio pulse is sent out by a cosmological event such that a range of radio frequencies are sent out at the same time. Here on Earth we listen in and detect each frequency in several bands. On Earth the pulse is detected and they arrive in order of frequency. The highest frequencies arrive first followed by lower and lower frequencies in succession. By measuring the time delay between the arrival times of two known frequencies we can determine the Dispersion Measure DM. The relationship between DM, n and distance d is:

$$DM = nd.$$

This is main stream Physics. Used all the time in radio astronomy and beyond criticism. It is correct.

This was the game changer in NTL. As soon as the DM of

a fast radio burst along with the redshift of the host galaxy was measured we were able to find the mean electron number density – which was exactly as predicted ten years earlier by NTL. As stated earlier, it was John and Louis who had brought my attention to this and how it could be used to find 'n.' Then I started to think about this and the rest is mine and mine alone!!

It struck me that Dispersion Measure was caused by photons interacting with the electrons in the plasma of IG space. There was an accepted relationship between DM and $'n'$ that worked and scientists involved in radio astronomy accepted it and used it as '*reliable.*'

I have a theory that says that redshifts are caused by the interaction of photons of light (or radio) with the electrons in the plasma of IG space.

The BB says that redshifts are caused by the stretching of space and are not related to the electrons in that space.

If I am correct then there should be a direct relationship between redshift and DM since they are both caused by the same effect – photons interacting with electrons in the plasma of IG space as they travel along.

If the BB is correct then the two effects, apart from a qualitative 'hand-waving' type of thing would not give a direct relationship.

The challenge was on!

From radio astronomy, we have $DM = nd$. From NTL we have

$$z = \exp(2nhrd/mc) - 1$$

rearranging this for d gives $d = \{(mc)\ln(1+z)\}/2nhr$

Since $DM = nd$ we can simply substitute for d to give:

$DM = \{(mc)\ln(1+z)\}/2hr$ which is a direct proportionality between DM and ln(1+z). That is $DM = \{mc/2nhr\}\ln(1+z)$. Consequently, a graph of DM against ln(1+z) should be a straight line through the origin with gradient $\{mc/2nhr\}$. The problem is (as ever) units. In NTL I use SI units. Unfortunately this is not the case in cosmology (and probably why they get their knickers in a twist so often). In the equation $DM = nd$, n has units cm^{-3} and d has units of parsec! Consequently to get a numerical value for the constant of proportionality we also need to include a unit conversion. When we do this and insert values for the Universal constants we arrive at:

$$DM = 2380\ln(1+z)$$

This is a prediction made by NTL and should be easy enough to test. Is a graph of DM versus ln(1+z) a straight line through the origin with gradient 2380 in which case NTL triumphs again? Or if there is no relationship – it fails!

What we need is data. We can't use the FRB from earlier as we have already used that to determine the mean electron number density and in any case one data point is not enough to construct a graph! Fortunately, in 2013, Thornton et al published redshifts and Dispersion Measures for a group of four pulsars and so we can apply our NTL predictions to these results. Now I know what you are thinking. Didn't we say that Keane et al had published redshifts and DM's for the first time in 2015? The answer is yes and we were correct as we have the DM from the Fast Radio Burst and the redshift is not from the object itself but from the host galaxy it was found in. Hence the redshift is a '*true*' one with no intrinsic redshift due to the object itself. Pulsars are known to have an intrinsic gravitational redshift which can be as large as 0.6 and this is added to the cosmological redshift to give the one measured at the Earth. So lets construct a graph of DM versus ln(1+z) and see if it is a straight line through the origin. We will plot the Keane data point for

comparison.

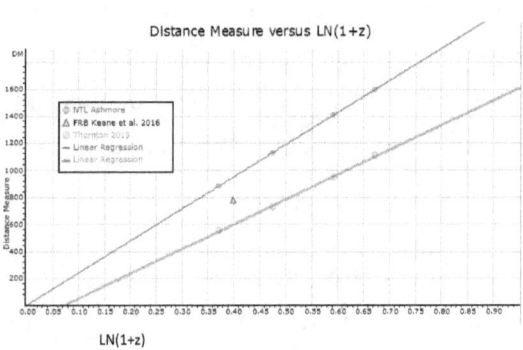

We see that the four data points form a linear graph. That said the line does not pass through the origin nor does the graph agree with the Keane data point. For those reading this book on kindle, the diagrams do not always come out right and so the originals can be found at:
http://file.scirp.org/pdf/JHEPGC_2016082515515600.pdf

Now, the graph does not pass through the origin because of the intrinsic gravitational redshift of the pulsars. A DM of zero means that the pulsars are zero distance away and yet there is a redshift. It must be due to gravity and not any cosmological effect. When we remove this intrinsic redshift so that the graph passes through the origin we are left with the cosmological component and we see straight away that there is good agreement between both sets of data.

All five points now agree with a direct proportionality (as predicted by NTL) between DM and ln(1+z). NTL predicts the relationship as $DM = 2380 ln(1 + z)$ where $2380\ m^{-2}$ is the gradient

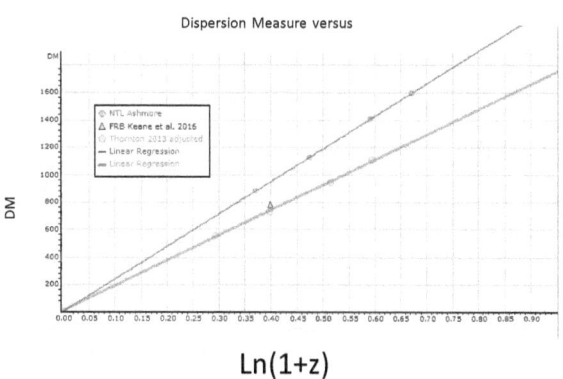

Ln(1+z)

The results from observation give $DM = 1830 ln(1 + z)$. Comparing the observation with NTL predicted.

i) Both give a straight line graph through the origin.
ii) The gradient from observation 1830 m^{-2} is close to the predicted gradient of 2380 m^{-2} and differ by only 23% which in cosmological uncertainties is good.

In the BB theory, DM is caused by photon–electron interactions whereby redshift is caused by the expansion of the Universe and so one would not really expect a direct relation other than by chance.

In NTL, both DM and redshift are due to photon-electron interactions and so we would expect a direct relationship. The observational points follow the same relation as that predicted and the gradients agree.

If, as in the BB these results are a remarkable coincidence then one must say that it is even more remarkable in that the predicted gradient is a combination of Universal constants. So let us ask BB supporters why does such a relationship occur and why is the gradient just about exactly equal to a combination of Universal constants few of which relate to expansion of the Universe.

New Tired Light QED?

BTW. Did you notice the switch in dispersion earlier? In light we have red light with the longest wavelength deviating the least when it enters a medium since it travels the fastest there. With radio we have longer wavelengths travelling the slowest! Now why? Why do the relative speeds switch over as we go through the e-m spectrum? Is there a wavelength between radio and light where they all travel at the same speed?

With sound this does not happen. All frequencies travel at the same speed through the same medium. No-one but no-one would go to an orchestral concert if sound behaved as light does. With light in glass, red with the longest wavelength travels the fastest' violet with the least wavelength travels the slowest. With sound, whilst all the instruments would play their notes melodically and in unison, their harmonies being beautiful in the first two rows, at the back of the hall it would be chaos with the bass notes arriving first and the high frequency piccolos notes arriving later. The notes would not arrive in the same order as they were played.

The reason of course is that sound is a vibration and the energy is passed from one molecule to the next without any delay. With e-m waves the speed depends upon

i) The number of photon/electron interactions there are.

ii) The time delay suffered at each interaction.

Red light has a longer wavelength than violet and so its collision cross-section is larger - it will make a larger number of interactions in travelling a certain distance. However, since it has a low energy the time delay suffered at each interaction is smaller than that of violet photons. In red light, the time delay outweighs the effect of number of collisions. Violet is slower as the photons take longer between absorption & re-emission.

In the radio, the wavelength is very much larger and so the collision cross-section is huge. Here the number of collisions outweighs the time delays. Lower wavelength photons arrive first since they have suffered much less collisions involving a greater time delay.

To put it another way, total time of transit, T is equal to the time taken for light to travel the distance d through the vacuum d/c, plus the total time of all the delays added together.

$$T = d/c + N\delta t$$

Where N is the total number of collisions and proportional to the wavelength whilst δt it the time delay between absorption and re-emission and δt increases with frequency or energy. For light, the relative wavelength and hence N is very small but the delay δt is high due to the relatively high energies compared to radio. For light δt is the dominant factor in determining which frequency travels the fastest. With radio, the wavelength and hence N is huge but δt is small due to the relatively low energies compared to light. For radio, the number of interactions N is the dominant factor. This is why we get the switch over in relative speeds as we go from light to radio frequencies.

Its just New Tired Light in action.

Chapter 20

Tests for NTL

When presenting NTL or any other theory for that matter at a conference, one of the first questions is 'Can it be tested in the laboratory?' This is a fair question and one which every theorist should ask themselves. Clearly, being able to calculate the redshift of any galaxy from first principles in terms of electrons (not to mention the CMBR) is great support for the NTL theory. Predicting the mean electron number density of IG space ten years before it was found is great support as is predicting the relationship between DM and redshift when no-one else expected it.

Before describing a test for NTL let us look at the problems in a laboratory test or indeed with Tired light theories in general. For a tired light theory one needs something that is everywhere in the Universe and evenly distributed since redshifts are basically the same whichever way we look. Redshifts are continuous so we need something that gives a small increase at each interaction. Most start off with a photon-hydrogen atom interaction (including myself but dropped early on) but I soon realised that the energy levels would cause havoc as photons would fall into them and like lemmings, never come out of the other side. The other problem with atoms is that if atoms were the cause, one would get a bigger redshift in our own atmosphere than one would from any cosmological effect. I know sunsets are red but this is a filtering effect rather than the shifting of spectral lines as in redshift. Characteristic lines tend to stay where they are in the spectrum even at sunset.

What we need is an effect that only occurs in the sparsely populated outback of IG space but not in our atmosphere and that is where plasma and recoil comes into its own. Firstly in IG space we are well away from the natural frequency of oscillation of the electrons and so absorption does not occur. All photons are re-emitted. Secondly, in IG space the plasma is '*squidgy*' which allows the electrons to recoil and redshifts to occur. As the plasma density increases electrostatic forces between the ions increases and these prevent the ions recoiling. They can still oscillate allowing for the transmission of light but redshift will disappear. For this reason it is possible to have a Dispersion Measure without a redshift (or at least a smaller redshift than expected). For dense plasma there will still be a delay between absorption and re-emission to give a DM but there will be no redshift since the electrons are unable to recoil – it is the whole plasma that recoils here and since it is very massive any effect will be negligible.

What I am really trying to say is that the obvious test is to pack a pipe full of high density plasma, fire a laser through it and look for redshift. One can't do this as when the plasma density increases the recoil stops and the redshift disappears! So in NTL, redshift z is proportional to electron number density n but only up to a point. As n increase eventually recoil and hence redshift will stop once the electrostatic forces between ions prevent them recoiling.

There have been reports of this being reproduced in the laboratory though. In 2009, Chen et al reported doing just the same.

_{https://www.researchgate.net/publication/228730621_Intrinsic_Plasma_Redshifts_Now_Reproduced_In_The_Laboratory-a_Discussion_in_Terms_of_New_Tired_Light}

Firing a high powered laser at a crystal target they reported redshifted emission lines as the electrons recombined with their original atoms. Importantly the redshift varied directly with the electron number density as predicted by NTL. The lines were also broadened (as all characteristic lines are) and this could well be due to statistical fluctuations in the number of photon-electron interactions. It must also be said that the redshift could be caused by the energy levels within the atoms changing due to the presence of the surrounding ions – so it's support and not proof.

Our Sun lends evidence in support of New Tired Light. The Sun is surrounded by plasma and there are reports of gamma rays emitted in solar flares are redshifted and the direction of the ejections rules out any Doppler effect. Interestingly there are reports that the wavelengths of characteristic lines at the edges of the Sun have a greater redshift than those from the middle (once Doppler rotation effects have been removed). Light from the edges travels further through the surrounding plasma so could this be a tired light effect?

Since we cannot increase the electron number density and reduce the distance to test for NTL maybe we could use the CMBR? This was the test that is said to have discriminated between the steady state theory and the BB theory so perhaps we could use that.

In NTL, the CMBR is the secondary radiation given off when UV, Light, IR etc. photons are redshifted. NTL correctly predicts the wavelengths of the resultant secondary radiation. Well the CMBR should also be redshifted. The peak of the CMBR is at a wavelength of $2.1 x 10^{-3} m$ and in NTL is caused by the interaction of a UV photon of wavelength $5 x 10^{-8} m$ with an electron. When this CMBR photon is redshifted a secondary photon of wavelength $3.64 x 10^6 m$ and frequency $82 Hz$ will be emitted. This is in the ELF range and is a frequency band used by submarines for communication. The technology is there to detect it. As a test of NTL, we need to look for a cosmic radiation, which is omnidirectional, non seasonal with a frequency distribution around 82Hz.

This would test NTL.

Chapter 21

Dark Matter

We have all heard of the mysterious dark matter and its effects on the future of the Universe but what exactly is it and better still can NTL cast any light on this dark matter? The answer is very illuminating to say the least. Firstly we need to decide which dark matter we are looking at and we will see that its our old friend Fritz Zwicky who starts the ball rolling.

In the early 1930's it had only just been realised that galaxies were groups of stars and Zwicky was studying a galaxy cluster. The Coma Cluster to be precise containing over 1000 galaxies. He had persuaded Caltech to build a special telescope for the job which would capture a large number of galaxies in a single image. From here he looked at their spectra and using the Doppler effect he measured their velocities. Since the velocities of orbiting objects depend upon gravitational forces acting on them he could determine the mass of the entire Coma cluster. He then measured the total light output from all the galaxies in the system which contains around 10^{12} stars. He then compared the ratio of mass to brightness for the cluster to that of a nearby stellar cluster and found that in order to account for the Newtonian gravitational motion there would need to be one hundred times more mass in the cluster than that provided by the stars he could see. Zwicky proposed that there must be a large amount of matter present that we could not see and named it *'Dark Matter.'* In the case of the Coma cluster the 'dark matter accounted for 99% of the mass of the whole cluster. This dark matter is real in that it comes from classical mechanics. Because of this Zwicky is known as *'the father of Dark Matter.'*

Next we have Vera Rubin who studied galaxy rotation curves. Before we move onto this just think about our solar system. As we go further from the Sun the gravitational field strength reduces and so planets at the edge of the solar system orbit much slower than those nearer the middle. It should be the same for galaxies as they rotate about the black hole in the middle. Stars furthest from the galactic centre should orbit with a lower velocity than those in the middle since the gravitational field is less.

Vera Rubin studied the radial velocities of stars as a function of their radius and found that this was not the case. The stars all rotated at the same velocity regardless of their distance from the galactic centre. It was as if they are all attached to a rigid rod which makes them all rotate together at the same speed. Either Newton's law of gravity does not apply on the large scale (which I do not believe for one moment) or it can also be explained by the galaxy being surrounded by a whole mass of stuff that we cannot see – dark matter.

I am putting this down as *'maybe real'* for two reasons. One, there is allegedly reports of this not happening within our own galaxy. That is when we measure the velocities of stars within our own galaxy one paper says the further out they are the slower they go – as per Mr Newton. This would imply that the rotational velocity problem is an *'outside'* effect. Two, I can imagine a galaxy rotating bunching up plasma at its outer arms. The further out one goes the more the plasma is 'bunched' giving you an increased measure red/blue shift and hence a greater velocity than that which happens in reality. This is just a thought but we will leave this as possibly real dark matter for now. Hopefully we will get more information on our own galaxy's stellar orbital motion to confirm the result from both inside and out!

Then we have the dark matter to make the BB work. This is mumbo jumbo and not a part of this book. Similarly for dark energy.

So how can NTL cast a ray of hope on this dark matter problem. Simple really. Dark matter is the plasma surrounding the galaxies and within them. Since plasma is transparent (light passes through it) we will not see it. Hence it is *'dark.'*

Think about glass. We only see glass because of the reflections from the front and back surfaces. If we were inside the glass we would not see it at all. Light passes through it with ease and with no visible effect. We would not know it was there.

It is the same when swimming underwater. Can you see the water? No. OK maybe the green chlorine in a swimming pool but in pure clean water you would not see it. It would be dark. Since most of the Universe it empty space it would not take a high plasma density to contribute 99% of the matter in a cluster. The only way we would notice it would be because of redshifts caused by NTL.

I have no issues with gravitational lensing but I dread to say that there is a possible NTL link. Gravity around a large mass would cause the plasma to form a sphere around it. Since the speed of light in plasma is less than in a vacuum. One would expect the plasma to thin as we go further away from the mass. The spherical shape thinning at the edges forms a perfect convex lens which would bend the light and magnify objects behind. You may well say that rotational effects would make the plasma into a disc along the plane of rotation. OK. We get two images one each side. One may say we would also gets jets of plasma perpendicular to the plane of rotation as well. OK that's four images so we will call it an Einstein cross. I think it was Halton Arp who said *'I can never understand why with gravitational lensing, the image is not a symmetrical circle.'*

Maybe this will have to wait till the next book!

Chapter 22

Is the Universe really cooling down?

We have all heard the story of the Hot Big Bang and how it has cooled down over the years so that all that remains is the CMBR – the '*echo*' of the BB. How the radiation has cooled down to such an extent that it is now at a temperature of 2.73K. We all know the shenanigans that went on in 'predicting' that number and this leaves little confidence in the result. Apart from the present day CMBR is there any other way of checking this?

Actually there is and the evidence is contained within the Lyman alpha forest and the Hydrogen cloud data. We have already looked at these absorption lines and used them to determine the spacing between Hydrogen clouds. We saw that for at least the last billion years or so they are evenly spaced showing a static Universe. But it is not just the spacing that gives us information on the past but the width of these lines is an important factor too.

The Doppler parameter, p gives an indication of the temperature of the Hydrogen clouds and is found from the width of the lines. An atom moving towards an incoming photon '*sees*' the photons frequency to be higher and so allows a lower energy photon to excite it. Similarly, an atom moving away from an oncoming photon '*sees*' the frequency as less and allows a photon of higher frequency to excite it. This gives us not just one frequency that will excite the atom and be absorbed but a range of frequencies and hence the absorption lines are broadened. The higher the temperature the greater the range of speeds of the atoms and so the broader the absorption lines. It must be said that the degree of disorder in the clouds also adds to the line broadening but the width will still give an upper limit to the cloud temperature. Data on the Doppler parameter exists in the form of Lyman alpha forest up to a redshift of around 3.2 and this gives us a look back time of approximately 10^{10} years. Since we have a dataset on temperature for the last ten thousand million years the we should be able to detect the cooling down of the Universe. Shouldn't we?

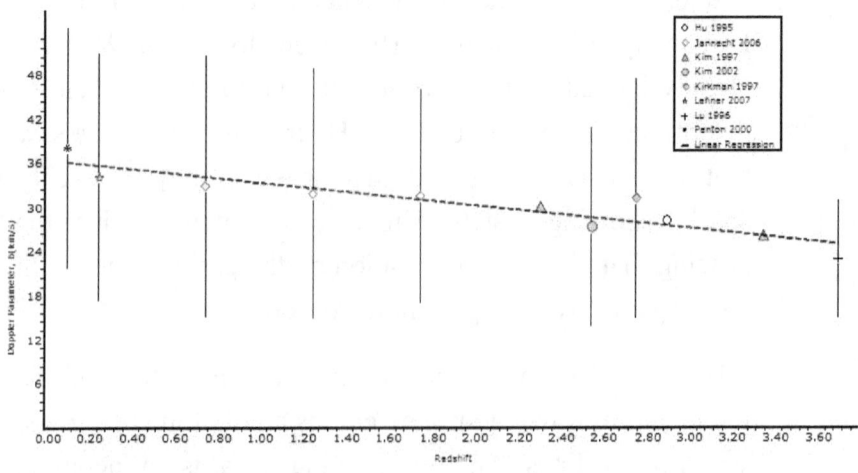

Well, no actually. If we look at the data over the last 10^{10} years the Universe certainly isn't cooling down but the temperature if anything, remains approximately constant or is even heating up! This must cast doubt upon the whole Big Bang Theory.

It gets worse for the BB when one looks at the Black Body nature of the CMBR. If we put our BB hats on for a moment the microwave background that we measure comes from different parts of the universe and it just happens that it all arrives at our detector at the same time to give a perfect black body curve. Now the radiation that came from further afield was emitted further back in time when the Universe was supposedly hotter and so had a shorter wavelength when emitted. As it travelled along it was stretched so that at the precise time of arrival it is now part of the black body curve at 2.73K. This piece of luck happens since as we go further back in time at increased redshifts the temperature increases and wavelength reduces as $(1 + z)$ but as the radiation travels towards us the wavelength stretches by $(1 + z)$ so the effects cancel. This is how all the radiation from differing parts of the Universe arrives with a black body spectrum of 2.73K and so when they all add together it remains black body.

However, with our Doppler parameter, as redshift increases the temperature remains the same (or at least the temperature of the Hydrogen clouds within that space). So as the radiation travels towards us it will stretch and no longer be at 2.73 on arrival but cooler. The black body curves would no longer be all for the same temperature and so when we superimpose them they would add to a spectrum that was not black body!

Unless the CMBR is local as in NTL.

Chapter 23

Hybrid Universe?

We are now certain that the Universe is static. Redshifts and the CMBR can be calculated from first principles, DM and redshifts have a predictable relationship between them and NTL gives us all the answers.

But did the Universe expand in the past?

The reason I ask this question is the Hydrogen cloud data. This shows that the Universe has been static for at least the last billion years or so, But what of beyond that? Taking the data a face value it would appear to show expansion in an early Universe as the clouds become less and less closely packed as the Universe aged.

We will let our imagination run wild for a moment and run with this idea and see if there is any way we can reconcile the Big Bang with Tired Light. That is, the Universe the Universe started with a Bang, expanded outwards but the density was at the critical density and so gravity arrested the motion and brought it all to an unstable but static end. Redshifts are now solely due to Tired Light since the Universe is now static. We can plot the size of the Universe from the Hydrogen cloud data. The 'size' would be directly proportional to the mean cloud spacing and we see that the Universe becomes 'static' at a redshift of approximately $z = 0.7$. Beyond this point the clouds show '*expansion?*' One might think that this could explain the supernovae results in that perhaps they are so far away that when the

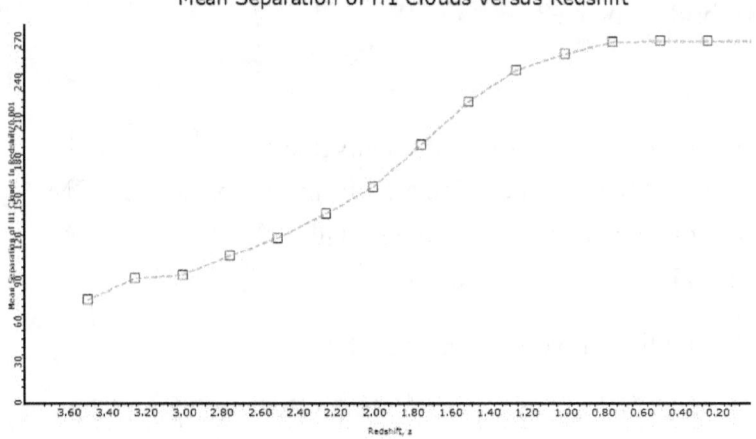

light was emitted the Universe was in its expanding state but this is not correct. The supernovae used for the claimed 'time dilation' are mostly at redshifts below $z = 0.6$ and so these will still be in the static region. No, time dilation is wrong whichever way one looks at it. Should that be '*spherically wrong?*'

This would have a great impact on age of the Universe. In present day BB the age of the Universe is the reciprocal of the Hubble constant but since we now know that H is a combination of Universal constants including the mass and radius of the electron then this is clearly not correct.

One way could be to take the H1 cloud data from the region that may show expansion and extrapolate back. The equation linking the number of clouds per unit redshift at a redshift of z (dN/dz) and the value now $(dN/dz)_0$ i.e the number of lines per unit redshift at $z = 0$ is:

$$(dN/dz) = (dN/dz)_0 (1 + z)^\gamma$$

Where γ is a constant and can be found from a graph of $Log(dN/dz)$ versus $Log(1 + z)$.

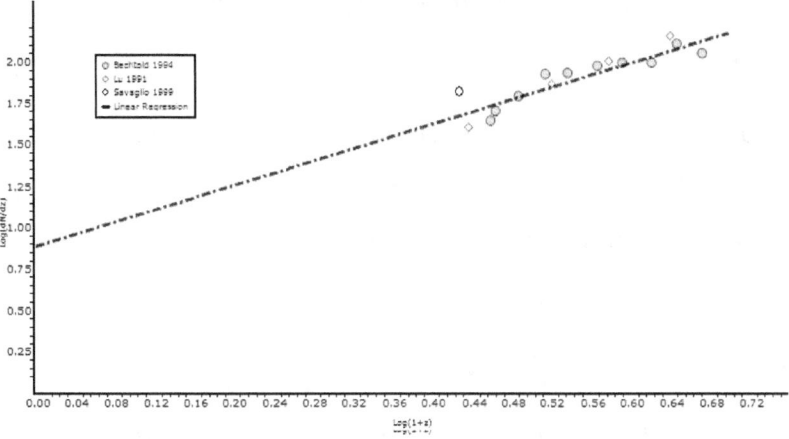

The graph tells us that γ has a value of 1.89 and $(dN/dz)_0 = 9$. So our relationship becomes:

$$(dN/dz) = 9(1 + z)^{1.8}$$

And we can use this to give some idea of how old the Universe is (at least in terms of redshifts).

Let's ask some questions.

The clouds have been moving apart supposedly as space expands and carrying them with it - so at what redshift were the clouds *'touching'* seems a good place to start i.e. H1 clouds are ≈ 70kpc in size, so there would be 6600 per unit redshift at the time that these clouds were *'touching.'*

Putting this in our equation shows that we would have to go back in time to a redshift of $z = 32$. Remember we said that a redshift of only 3.2 was 10,000,000,000 years so how old is this – and this is only to fill in the gaps and bring the clouds together.

We could go further and ask the question, "at what redshift did the clouds have an 'atomic separation' of $10^{-10} m$?" Here there would be $1.5 x 10^{36}$ H1 clouds per unit redshift. We would have to go back in time to a redshift of $z = 4.32 x 10^{18}$.

It is clear that if this is the case (and remember we are just looking at one set of observations so it may well be that the Universe has been static all its life) the Universe is so old it could be thought of as infinite in terms of us mere mortals.

Chapter 24

A good place to stop.

This seems a good place to stop – especially as I have just received an email from academia.edu saying I am in the top 5% of researchers there in terms of 30 day views of published papers. Since there are over 47 million researchers that use the site as a repository its not bad and shows just how much interest there is in Tired Light.

In many ways this seems the end of a journey (don't worry it won't be) in that I have completed many of the things I first set out to do and have had the results published in peer reviewed journals. This is the reason for this book on Tired Light (or as Sherlock Holmes would say *'a monograph on Tired Light'*).

Why now? Mainly the discovery of the FRB and the measurement of both the DM and the redshift of the host galaxy was a first time event and enabled a value for the mean electron density to be found from observation. I had been saying it was $0.5m^{-3}$ for over 20 years and first published this in 2006 - a fact that can be checked! This and the realisation that Dispersion Measure is a new area in which to apply 'Tired Light' and to show it works here too really brings us around full circle.

How did I get into this is probably a good question to finish with and the answer lies back in 1995. The Big Bang Theory had just been included in the Physics specifications and as a Physics teacher I wanted to know more about it. I had studied Physics at York University and well remember the astrophysics course and the accompanying text by Sir James Jeans on 'Astronomy and Cosmogony' which was just about as exciting as *'Statistical Thermodynamics I'* and if that wasn't bad enough it was later followed by '2' and '3' to rub it in!

As a Physics teacher I was working Dubai, UAE having gone to the Middle East for one year , two years maximum and at the time of writing I am still here after 34 years. Tuesday morning, 1995, module 1, Year 11 lesson on the expanding Universe... *'and as the Universe expands it increases the wavelength of the light so that it is now nearer the red end of the spectrum and gives us redshift...'* End of lesson. The class is saved by the bell and off they troop;

module 2 and in walks Year 12 A Level Physics for their lesson on one of the topics included at the time. This option was medical Physics....' Of I spout again; '*if the energy of the X-ray photons is high enough then they will suffer Compton scatter. Its energy and frequency are reduced and the wavelength increased...*'

It is at moments like this that as a teacher you stop to think. Why am I using the stretching of space to explain an increased wavelength in module 1 and a photon loss of energy in module 2? Moreover, if in module 2 the photon lost energy as its wavelength increased then why did it not lose energy in module 1 with the stretching of space? I put the thought to one side and went back into teacher autopilot. A few weeks later it was the school summer break and we as a family went to Cyprus for the duration apart from a two week 'duty trip' to the UK to see relations.

Sat on the balcony with a bottle of duty free Bacardi to one side and a couple of cardboard cartons of local red wine at the other, wife and kids gone into town I watched the Sun go down in style and as it went down it became redder and redder. This is a different effect than redshift as it is due to the atmosphere filtering out the blue but it brought back the thoughts of earlier in that could redshift be caused by the intervening medium? I thought back to those two back-to-back lessons and wondered about Compton scatter with the Hydrogen atom being the culprit and determined to look into it further. I tossed several ideas around for the next few days and realised that it could not be Compton scatter as here the photon went off in a different direction and I knew well that light travelled in perfect straight lines.

I did have some background on this as I had done a Master of Philosophy research degree at what was then the Preston Polytechnic and despite the grand name of the thesis it was really to do with developing solar cells. With a solar cell the trick is to get the photon absorbed in the depletion layer whereas what I wanted here was the photon not to be absorbed totally but to be re-emitted.

Once in London I hit the libraries of Imperial College and the British Library. In those days one could walk into any University Library just by signing in and no-one bothered. I seem to remember signing in as Mickey Mouse once or twice. It was the same with the British Library, I just signed up and was given a ten year library card that enabled me to call up any paper I wanted. Not so these days as access seems to be restricted to people linked to an institution and independent researchers such as myself seem to be barred. Is this censorship? Back in the UAE there was a great University Library in Al Ain but it involved a day trip to cross 100km of desert to get there. In any case, with the Internet the libraries are more or less redundant and any research can be done from home.

It was in London that I came across the Mossbauer effect upon which NTL is loosely based. The Mossbauer effect involves gamma rays emitted from the nucleus of an atom as it falls from a higher energy level to a lower one. In principle we should be able to get resonance absorption whereby a second identical nucleus absorbs the original photon emitted by the first as it goes from a lower energy level to a higher one. This does work in solids but not with gases and the reason is recoil. In a solid the nucleus is fixed in an atom which is fixed in a crystalline lattice and so it is the whole crystal structure that recoils and since this mass is huge on the atomic scale the recoil effect is negligible and there is no energy loss to the emitted photon.

With a gas this is not the case. The nucleus and atom recoil so that the emitted photon does not have as much energy as before since some of it has gone to the recoiling nucleus. The energy of the photon is no longer equal to the difference in energy levels in the second nucleus and so is unable to excite it. The Mossbauer effect is tested by moving the emitting nucleus at various speeds to counteract the recoil and until resonance absorption takes place again.

Here I had my effect. The advantage being that recoil takes place along the line of sight and so the photons would continue in a straight line. It had another bonus in that it explained why our atmosphere did not effect the redshift. It is a simple calculation to find the '*optical depth*' or what distance in space is our own atmosphere equivalent to. A sort of $n_1 d_1 = n_2 d_2$ with n_1 and n_2 being the number density of our atmosphere and space respectively, d_1 the thickness of our atmosphere and d_2 the equivalent distance in space and it turns out that our own atmosphere would produce a redshift which would drown out most of those in space! Our atmosphere is equivalent to around $6 x 10^{15}$ light year of IG space if you are interested. The idea with Mossbauer that recoil stops when the density increases overcomes this problem.

Back to Cyprus and whilst most people enjoyed the Sun and beach at Curium in 1995, I joined in but continued to work on the Mossbauer effect and how to apply it to the cosmos. This is the good thing about theoretical Physics – it can be done on the beach with pint in hand. I had calculated the increase in wavelength due to recoil and knew it was independent of wavelength. That meant, I reasoned it had to be the collision cross-section that had the wavelength dependent term in it as $z = \Delta\lambda/\lambda$ and longer wavelengths must have longer increases in wavelength so that the ratio remains constant for all wavelengths. In any case, it seemed sensible from diffraction effects. As collision cross-sections are areas I reasoned they must have something to do with the particle size and so I came up with the collision cross-section as the diameter of the Hydrogen atom multiplied by the wavelength or $\sigma = 2r\lambda$. At this time I had no reason for doing this other than it seemed about right and even with Hydrogen atoms I was getting numbers which were just about right.

At this time, I was just about boring everyone with my Tired Light theory including my long suffering son Ruben (who would listen as long as I kept buying the pints – he was 16 and this is Cyprus!) and during one of these evenings when I was updating him on latest developments it occurred to me that I could test the theory by calculating the Hubble constant. We were in 'Pot Black' at the time, a snooker club and bar we used to frequent in Limassol and a quick mental 'order of magnitude' calculation in my head confirmed I wasn't far off. Being in the right place for a celebration, we did.

Over the next few days, I typed it up and sent it off for publication. Within five weeks, start to finish in the Summer of 1995, I had the basic theory including a calculation of the

Hubble constant. Nature, ApJ and A&A being top of the list and by return of post I received three rejections. I still have them.

It may be worth mentioning that at this time I had never heard of Zwicky and tired light theories or even Halton Arp and other dissident cosmologists. It just seemed that the best explanation of these redshifts was a photon loss in energy to matter. J.J. Thompson summed it up best when he said that when embarking upon a new area of research, one should not read anything about it until you have developed your own ideas. Reading about it can make you follow other peoples train of thoughts and set you off on the wrong path.

It is difficult to drop a theory or idea when it is giving you the correct results (or at least close to a correct result) but I was still not happy about the Hydrogen atom. If the Hydrogen atom could produce redshifts then so too could every other atom in our atmosphere and I was not convinced that these atoms couldn't recoil as was required to stop them producing redshifts of their own. I was puzzling over this on leaving work and as I walked across the car park I realised that energy levels killed the theory – or at least the Hydrogen atom as culprit.

With redshift, absorption lines must move across the

spectrum towards the red end. The energy of the photons would reduce but when they reached an energy level they would be completely absorbed by resonance absorption. Black absorption lines should sweep across the spectrum as they are redshifted. Photons suddenly couldn't just reappear at the other side as they had gone forever. The absorption lies caused by the Hydrogen atom itself in IG space would just get broader and broader which does not happen. Once I realised this, out went the Hydrogen atom and I replaced it with electrons in the plasma of space. That night I rewrote the paper, recalculated and this time the numbers came out much nearer. Whilst the exact number density was not known it was thought to be one atom of Hydrogen per cubic metre of space and this is the value I was using giving me double the required values. But there was no actual measured value until much later (2015). Electron in the plasma of IG space have a resonant frequency of around 30Hz and so all our *'light'* photons are well above this. Resonant absorption is no longer a problem and now that we have charged particles, long range electrostatic forces would prevent recoil in dense plasma. Job done.

In the rejection slip, the editor of A&A did say that he wouldn't publish the theory until I had published collision cross-sections - which was fair enough and so I spent the next several years scouring libraries for published values as these were the only place to look. As the internet grew I would search the web for key words like *'photon'*, *'electron'* and *'cross-section'* but to no avail. One day I stayed late at work as I was running an astronomy evening and it wasn't worth going home between the end of school and the arrival of the dark skies needed for observations and so I went back to the internet for an hour to continue my search for the illusive σ. I had tried the usual key words and as many other words that might be relevant when I had a bright idea. I knew that the theory worked with $\sigma = 2r\lambda$ so why don't I just put that into the search engine and see what comes up? Instantly it was there on the screen, *'$2r\lambda$ - photo-absorption cross-section for low energy X-ray photons interacting with matter. Lawrence Berkeley National Laboratory.'* I had it! One of the Authors was named Hubble so that had to be a sign hadn't it? On my next trip to London I visited the British Library and photocopied hundreds of pages of the data for the various energies and found it really was $\sigma = 2r\lambda$. Back to Dubai, I rewrote the papers and sent it off to A&A in Paris (well he did say he wouldn't publish it without published cross-sections and these I now had). I received the reply to the effect that we now only publish papers by people in scientific institutions!

ApJ rejected it with something like 'don't send us anymore papers!'

It was at this point I started to think about the CMBR and if I could include this in the theory. You are probably wondering if I have spent the last 20 years doing nothing but think about Tired Light and you are probably right! Holidays, dinner parties and everything else were spent working with Tired Light. Someone once asked Isaac Newton how he had come up with his theory of Gravity and he answered *'by thinking about it all the time.'*

I was invigilating an examination – something that has to be the most boring thing ever - the only excitement being a student requiring an additional sheet of paper which produces a scurry of activity as all the invigilators wanted the excuse to do something; anything! It was here that I was toying with the idea of the energy transferred by recoil being re-emitted as secondary radiation which would form the CMBR and realised it could be done. But did I want to do it? Having spent many years on this theory and invested a lot of time what would happen if it came out wrong? Would I have to abandon the whole project? But if it came out correct…. I did a quick order of magnitude calculation in my head and saw the predicted wavelength was about right and left the examination room probably the happiest person there!

The last piece of the jigsaw was introducing the non constant collision cross-section. Up till this point I was considering it constant and equal to the initial value. This is fine for relatively nearby galaxies – we used it here in the *'from first principle'* examples and it gave us good results. But for completeness the theory had to incorporate the increasing wavelength so it could cope with far off galaxies. As the photon travels along it interacts with electrons and is redshifted. This increases its collision cross-section so it does not travel as far, on average, before it interacts again. I tried several ways of incorporating this summing algebraic series and all sorts but to no avail. During the summer holidays in Acapulco, Mexico (again – as a teacher it is hard to get anything of your own done in term time as life is too frenetic at school) I put pen to paper each morning and sorted out the calculations. I vaguely remembered something from far back that if one had a large number of terms with small differences between them then the sum could be found by integration; and that was how it out came - the exponential form of the Hubble equation. It was only later that I found out about Zwicky who had come up with the same result eighty years earlier! Looking back its hard to understand why all this took so long but I don't claim to be a genius - just stubborn and wont give up. Mind you I did spend time being serenaded on the beach by mariachis (yes just me) and quite a while in the cliff-top bars (Barbarossa's in particular).

Not having had any success at getting the theory published I decided to write the book, 'Big Bang Blasted!'. This was one of the best decisions ever as not only did I thoroughly enjoyed writing it but I still get emails from people telling me how much they enjoyed reading it. That book told the full Big Bang story including the characters involved and dispelling some of the myths built up around them. With this book I just wanted to tell the story of New Tired Light.

As soon as the book came out the paper was accepted published by Galilean Electrodynamics! Since then my papers have been published by Astronomical Society of the Pacific, appeared in ArXiv, and several conference proceedings. Otherwise it has been pretty quiet up till recently.

I spent quite a bit of time posting on internet forums to get the idea of tired light out until in the end I was banned from most of them. Not for anything sinister, just that the moderators don't like posters being right and the discussion continuing too long. It was at this time that I introduced 'Ashmore's Paradox' to wind people up. It turns out that in SI units the Hubble constant is numerically equal to hr/m *per cubic metre of space*. This comes out as $2.1x10^{-18}\ s^{-1}$ *or* $64\ km/s\ per\ Mpc$. The reason behind this was I had been teaching somebody's law or other when a student stuck his hand up and asked, '*Is there an Ashmore's Law, Sir.*' I thought why not a paradox – sounds more classy.

So what changed now? For the first time the DM and redshift of a FRB was measured and that changed everything as now we had an actual value for the mean electron number density and I went back to work on NTL.

With the new found value for n I went on to show that in NTL, since Dispersion and redshift are both caused by photons interacting with electrons in IG space there should be a relation between these quantities. I calculated the relationship, found a set of data on DM and redshift and showed NTL had it right. The relationship predicted by NTL was correct. In the BB DM's are due to electron interaction, redshift due to a stretching of space and so any relation would be pure fluke.

Excited! Now what to do. I had intended to submit a paper to a conference in the Summer of 2016 in Italy and go and present it but since I was moving home half-way across the World at the time I didn't think my wife would let me walk out and leave her to do all the moving. So that is how I came to submit the paper to a peer reviewed journal. I had not long before been asked to review a paper submitted to the *'Journal of High Energy Physics, Gravitation and Cosmology'* or *'JHEPGC'* for short so I thought, *'why not?'* The paper was short and included the number density, a reminder that I had predicted this value time and time again for the last ten years in copyrighted journals along with the relationship between DM and redshift – which I was quite proud of. The paper was originally about three pages long but nevertheless I was satisfied by it. Probably politically incorrect to say so but by the time the reviewers and adjudicators had finished it had grown to four times that size. I have to say how grateful I am to the editor of JHEPGC and these people for taking the time to recommend improvements which I included with gratitude. In the end the paper is a complete summing up of NTL. But let me digress for a minute.

I wouldn't say a physics discovery is better than sex but it certainly lasts longer. I have lived with NTL for over 20 years and it has been and continues to be the ride of my life. I recently read an interview in which J. K. Rowling was asked about Harry Potter. I have probably got this wrong but she said something along the lines *'I lived with Harry Potter for seven years when no-body else knew of him.'* I know how she felt (not about the money!) but having this secret about which you are the only one in the World to know). I suddenly realised that these two separate parts of astrophysics should be related if I was to be correct. I worked out the relation, found the data and spot on!

Something in me didn't want to share it. Stupid I know but it has happened several times on my journey. Two days was enough for me for DM versus redshift to be my own private property so off it went.

The rest is history.

http://www.scirp.org/Journal/PaperInformation.aspx?PaperID=70089

Appendix

Fritz Zwicky

In 1929, Zwicky published his thoughts on a Tired Light theory. Titled *"On the Redshift of Spectra Lines Through Interstellar Space"* and published in the *'Proceedings of the National Academy of Sciences'* he set out his ideas on what he described as *'a sort of gravitational analogy of the Compton effect.'*

Redshifts of distant galaxies were new at that time; indeed Zwicky still referred to galaxies as *'nebula'* and it had only been four years earlier when Hubble had measured the distance to the Andromeda galaxy settling the dispute that there was more out there than our own milky way galaxy – that is, that the universe consisted of billions of separate galaxies and not just the one – our milky way. Zwicky's paper was published in the same year that Hubble's presented his data claiming; *'the data in the table indicate a linear correlation between distances and velocities (of nebula and hence galaxies),* leading to the Hubble Law which tells us that a galaxy twice as far away will be travelling away from us at twice the speed (v = Hd). Zwicky was either a fast writer of scientific papers or he was using private data gleaned from the weekly meetings held in Hubble's front room where the researchers met to discuss the results and thoughts of the week.

He began his paper with a review of the evidence existing at the time and labelled the points A, B and so on. We will do the same here as we discuss the paper.

   A. Introduction. – *"It is known that very distant nebulae, probably galactic systems like our own, show remarkably high receding velocities whose magnitude increases with distance. This curious phenomenon promises to provide some important clues for the future development of our cosmological views. It may be of advantage therefore, to point out some of the principal facts which any cosmological theory will have to account for. Then a brief discussion will be given of different theoretical suggestions related to the above effect. Finally, a new effect of masses upon light will be suggested which is a sort of gravitational analogy of the Compton effect."*

B. Discussion of the observational facts.

(1) Hubble has recently shown that the further away a galaxy, the greater its recessional velocity and that the relation is roughly linear. There are some large deviations to the rule for nearby galaxies which may be due to their peculiar velocities (the motion of the galaxy itself through space). Observations by M. Humason indicate that for very large distances the redshifts are so great that they cannot be attributed to the peculiar motions of individual stars within the galaxy but must be *'accounted for in some another way.'*

(2) The ratio of the shift in frequency (Δv) to the frequency (v) represents the velocity and this ratio is independent of the frequency of the spectral line. However the available range in the spectrum is not very large. There are also some exceptions that have been found which suggest that the ratio for the Hydrogen line in the gamma (H-gamma) is slightly different than

(3) the Hydrogen line in the blue (H-beta).

(4) The lines themselves are just 'shifted' and there is no appreciable change in the appearance of the lines. Therefore this effect cannot be attributed to absorption or scattering processes.

(5) The optical image of the galaxies seems to be well defined with no blurring. Only the distance to the galaxy is involved in the redshift process.

(6) The spectral lines obtained from these galaxies are not well defined and no systematic study of their shape has been carried out. M. Humason gives the width of these lines as between 40×10^{-10} and 70×10^{-10} m for two of the Andromeda galaxies.

(7) If the Hubble relation is applied to objects in our own galaxy then the effect would be too small to be noticeable. *"The theoretical considerations proposed by the author in the following*

(Zwicky's tired light theory) made it probable that an appreciable effect should also be observed in our galaxy." He goes on to say that this prediction was tested by a Dr. Bruggencate and showed that the velocity of recession of globular clusters does indeed increase as one looks through the galactic equator and thus the light travels through a greater amount of galactic matter.

In summary; what was known at the time was that the ratio of change in frequency divided by the frequency ($\Delta v/v$) at that time associated with the recessional velocity of the galaxy increased in direct proportion to its distance from us. There is no change in the lines themselves due to scatter of absorption and there is no 'blurring' of the image of the galaxy - the effect depends only on the distance. Zwicky is proposing a gravitational effect that would explain this shift in frequency without resorting to velocities. However this would mean that there should be an effect due to the masses residing in our own galaxy and he is comforted by a colleague who has reported a greater shift in frequency for light travelling through the galactic plane and hence past a large number of masses compared to that which has travelled the comparatively short distance perpendicular to this plane.

He then goes on to look at the different theoretical possibilities of accounting for these phenomenon.

C. de Sitter's Universe. – *"de Sitter has pointed out that the special type of space proposed by himself as representing our universe would imply on the average velocity of recession of the far distant nebula. But the linear relation of Hubble's can only be obtained by making some additional assumptions about the distribution of the nebulae."* In explanation, Zwicky refers to a recent paper by R. C. Tolman. Here Tolman explains that to explain the redshifts in de Sitter's model of the universe, galaxies had to be continually entering and leaving the observation range leading to a Doppler effect – this tends to be positive and increases with distance but to achieve the exact Hubble result requires so many restraints to be put on their orbits that it would be highly unlikely. Zwicky agreed that the de Sitter model could account to points B1 – B5 above but felt that to explain Bruggencate's results of an effect in our own galaxy would be *'unsurmountable'* and therefore rejects this model as an explanation of redshifts.

D. The Compton-Doppler Effect of Free Electrons.

It was known that there was an 'adequate number of free electrons,' in intergalactic space. Observations of lines due to ionised calcium and Sodium showed this and so *'one might expect light coming from distant galaxies would undergo a shift to the red by Compton effect on those free electrons.'* Zwicky then carried out some order of magnitude calculations and showed that the effect was so small in a single event that a great number of collisions would be needed to produce the size of the redshifts measured. Since in Compton scatter the photons of light go off in some other directions Zwicky concluded that this would make *'interstellar space intolerably opaque which disposes of the above explanation.*' He conceded that the electrons could have very high speeds as suggested by the existence of cosmic radiation and thus the shift in frequency could be done by one collision but he still felt that the scattered radiation would be a problem – especially since the images were of such good definition and not blurred.

E The Usual Gravitational Shift of Spectral Lines.

It was known that gravity itself can produce redshifts. Imagine a photon of light travelling vertically upwards from a mass. The photon gains Gravitational Potential energy and this has been transferred from its own energy. Since energy E = hf, if the photon loses energy the frequency reduces and the wavelength increases. It is redshifted. Zwicky considered photons travelling from one point in a galaxy to another. The nearer the source to the galactic centre the more gravitational Potential energy it would gain as it left and the greater the redshift. However, as he pointed out, this effect has no relation to the distance between one galaxy and another and so could not explain the redshift – distance law.

E. The Gravitational "Drag" of Light. Here Zwicky sets out his Tired Light Theory. According to relativity, a photon of light has inertia and a gravitational mass ($h\nu/c^2$) and so he proposed that the light would interact with masses in the universe. As the photon moves alongside a stationary mass gravitational forces would cause it to recoil. The mass gains some energy from the photon and so the energy of the photon is reduced, its frequency decreased and its wavelength increased. It has been redshifted. Furthermore, the greater the distance travelled by the photon, the more masses it will pass and thus the greater the redshift. He states that to get a completely satisfactory theory would require the use of the general theory of relativity but gives a *'rough idea'* by considering an actual small mass travelling along the x axis and causing a large mass set to one side to recoil. If the small mass is travelling fast enough then Zwicky argues that its deflection would be small. He then replaces expressions derived for the small mass with relations ships for the photon. An order of magnitude calculation shows that it gives values of redshift comparable to measured values – including those of globular clusters around our own galactic disc as measure by Dr. Bruggencate.

Zwicky, F. 1929. *On the Red Shift of Spectral Lines through Interstellar Space.* PNAS **15**:773-77

www.ingramcontent.com/pod-product-compliance
Lightning Source LLC
Chambersburg PA
CBHW070322190526
45169CB00005B/1699